Edward Burnett Voorhees

First Principles of Agriculture

Edward Burnett Voorhees

First Principles of Agriculture

ISBN/EAN: 9783337312596

Printed in Europe, USA, Canada, Australia, Japan

Cover: Foto ©berggeist007 / pixelio.de

More available books at **www.hansebooks.com**

FIRST PRINCIPLES

OF

AGRICULTURE.

BY

EDWARD B. VOORHEES, A.M.

DIRECTOR OF THE NEW JERSEY AGRICULTURAL EXPERIMENT STATION,
AND PROFESSOR OF AGRICULTURE IN RUTGERS COLLEGE,
NEW BRUNSWICK, N.J.

SILVER, BURDETT AND COMPANY.
NEW YORK. BOSTON. CHICAGO.

PREFACE.

THE purpose of this book is to state in logical order the elementary principles of scientific agriculture, and to show the relation of these scientific facts to farm practice. The need of such a book has been strongly felt by the author in his work as a teacher, not only of college students, but of those already engaged in farming.

His experience, both as a practical farmer and as a teacher of the theory and application of agricultural science, leads him to believe that the principles and relations of scientific agriculture, if set forth clearly and in a scientific manner, can be successfully taught in our country district schools. It is here that such education must begin, if it is to reach and influence the mass of farmers, upon whom rests the chief burden of irrational practice, and through whom must come any direct progress in the true development of the farming industry.

In the preparation of the book no attempt has been made to cover the whole field of the various sciences in their relations to agriculture; in many branches, only simple facts are stated, though it has been the aim to make the book scientific as far as it goes, and thus a safe guide to practical men in their farming operations, as well as a foundation upon which further study may be based.

CONTENTS.

CHAPTER		PAGE
I.	THE CONSTITUENTS OF PLANTS	7
II.	ORIGIN AND FORMATION OF SOILS	18
III.	COMPOSITION OF SOILS	29
IV.	THE IMPROVEMENT OF SOILS	41
V.	NATURAL MANURES	52
VI.	ARTIFICIAL AND CONCENTRATED MANURES; NITROGENOUS MATERIALS	68
VII.	ARTIFICIAL AND CONCENTRATED MANURES; PHOSPHATES	80
VIII.	ARTIFICIAL AND CONCENTRATED MANURES; SUPERPHOSPHATES AND POTASH SALTS	90
IX.	ARTIFICIAL MANURES OR FERTILIZERS; METHODS OF BUYING; VALUATION; FORMULAS	101
X.	THE ROTATION OF CROPS	113
XI.	THE SELECTION OF SEED; FARM CROPS AND THEIR CLASSIFICATION; CEREALS; GRASSES; PASTURES; ROOTS; TUBERS; AND MARKET-GARDEN CROPS	122
XII.	THE GROWTH OF ANIMALS; THE CONSTITUENTS OF ANIMALS AND ANIMAL FOOD; THE CHARACTER AND COMPOSITION OF FODDERS AND FEEDS,	137

CHAPTER		PAGE
XIII.	THE DIGESTIBILITY OF FODDERS AND FEEDS; FEEDING STANDARDS; NUTRITIVE RATIO; THE EXCHANGE OF FARM PRODUCTS FOR CONCENTRATED FEEDS	154
XIV.	PRINCIPLES OF BREEDING; THE PURE BREEDS OF FARM STOCK	170
XV.	THE PRODUCTS OF THE DAIRY; THEIR CHARACTER AND COMPOSITION; DAIRY MANAGEMENT	182

APPENDIX —

TABLES 199

INDEX 209

FIRST PRINCIPLES OF AGRICULTURE.

CHAPTER I.

The Constituents of Plants; Plant Growth.

Parts of Plants. — Most agricultural plants possess three distinct parts, — the root, the stem, and the leaf. The main uses of the root are to secure food from the soil, and to serve as a support to the plant during its period of life. The stem acts as a support for the leaves, as a medium for the circulation of food through the plant, from the leaf to the root and from the root to the leaf, and as a storehouse of nutriment for future growth. The leaves secure food from the atmosphere, and permit the escape into the air of the water taken up by the roots.

These different parts all co-operate, or work together, to secure and distribute the constituents necessary to the complete growth and development of the whole plant. What is plant food, or of what constituents is a plant composed, is a question of the first importance in a study of the growth of plants.

The Water contained in Plants. — A plant in the first place is composed of two distinct classes of substances, — water and dry matter. Water is contained in all growing plants; forest and fruit trees seldom contain less than four-fifths, while vegetables and young plants are frequently nine-tenths water.

When plants are removed from the soil, and brought in contact with the air and warmth, a large part of the water contained in them escapes in the form of water vapor; they become what is termed "air dry." The proportion of water lost in this way, and the time or rapidity of loss, depend upon the moisture contained in them, and the warmth of the atmosphere; the drier and warmer the atmosphere, the greater the loss of water. This loss of water is nicely illustrated in hay-making, the time required to dry or cure depending upon the kind of plant and the character of the weather.

Air-dry Plants. — In temperate climates, air-dry plants still contain from eight to twelve per cent of water. To effect its complete removal they are heated to a temperature of 212° F., that of boiling water, until there is no further loss of weight. The portion remaining after the drying is termed the "dry matter." The dry matter of plants contains all the constituents of their growth; that is, all those necessary for perfect growth, no one of which can be removed without destroying it.

Water, while it is essential in the growth of plants and serves a useful purpose, is not a constituent in the same sense as those which are contained in the dry matter, since, as has already been noted, it can be removed without destroying the structure and form of the plant.

The Total Dry Matter. — The total dry matter of plants contains all those substances or compounds which are useful for the purposes for which they are grown. The constituents contained in it may also, for convenience of study, be divided into two classes: first, those which are lost or driven away by burning, or are capable of being burned; and second, those which are contained in the ash or residue after burning. The first are called "combustible" or burnable constituents; the others are called the "incombustible" or ash constituents.

The part that is removed by burning contains the constituents that have been derived largely from the air, while the remainder contains the mineral substances which have been derived largely from the soil. The burning of wood illustrates the separation of these two classes of constituents and the proportions of each.

Each one of the constituents contained in a plant is a distinct chemical element; and the number and kind which have been found to be absolutely essential to the growth of plants are carbon, oxygen, nitrogen, hydrogen, potassium, magnesium, calcium, iron, phosphorus, and sulphur. These are called plant-food elements, and all healthy plants contain them; if, by any chance, any one of them could not be obtained, the plant could not make normal growth, and in this sense no one of them is of more importance than another.

Besides these, however, the chemical elements, silicon, chlorine, sodium, manganese, and sometimes others, do exist in the plant, though plants can be grown without them. Thus, at most, but fourteen only of the seventy known chemical elements are necessary for the growth of a plant, and form its food. The large number of different species and varieties of plants now existing is, however, capable of being produced from them, the different properties and forms being entirely due to a different grouping or combining of the constituents in the plant.

Source of Plant-food Elements. — Plants derive the elements of their growth from two sources, — the atmosphere and the soil. The atmosphere, while the original source of carbon, hydrogen, oxygen, and nitrogen, furnishes direct chiefly the element carbon. Hydrogen and oxygen are furnished mainly through the soil in the form of water; though the original source is the atmosphere, where they exist in the form of "water vapor." Nitrogen is also largely taken up by the roots of plants; though certain classes of plants, as we shall see later, have the power, under proper conditions, of obtaining it from the atmosphere, of which it constitutes about four-fifths of the whole bulk.

The Proportion of Food derived from the Air. — These constituents derived from the atmosphere constitute about ninety-five per cent of the total dry matter of plants, of which carbon constitutes nearly one-half; the remaining five per cent is the "incombustible" or ash constituents. The soil is the exclusive source of these elements; they are, however, indispensable, for without them the carbon of the atmosphere, the hydrogen and oxygen of the water, and the nitrogen of the soil or atmosphere, would not have been able to enter into plant life.

The atmospheric constituents are usually termed the "organic;" and the soil constituents, the "inorganic," or ash elements. This distinction, though not entirely accurate, serves a good purpose in helping to get a proper understanding of the relations of the compounds formed in the plant.

The Essential Constituents, How Determined. — The sources of the chemical constituents necessary to the growth of plants, as well as their kind and number, have been determined by careful experiments, conducted in the following manner: —

Sand, which is an inert substance, is thoroughly burned, which destroys all combustible matter; then carefully washed, which removes all traces of plant-food; a portion is then put in a suitable jar or box, and seed of wheat or corn planted, and moistened with pure distilled water. The sand contains no food, yet the weight of the dry matter in the growth made is shown, by careful analysis of both seed and plant, to be much greater than the weight of the seed planted; also, that the increase in weight consists entirely of carbon, hydrogen, and oxygen, which must have been obtained from the air and water. This establishes the fact that air and water furnish food, and that this food consists of atmospheric elements.

In another series of experiments, to the sand and water, in one experiment, the various essential ash or mineral constituents alone, are added; to another, nitrogen alone; while to a third, both the ash constituents and nitrogen are added. In the first and second experiments, but little addition is made in the weight of the crop over that secured when water alone was added; the growths are not perfect; minerals alone, and nitrogen alone, are not sufficient. In the third experiment, however, the crop is fully developed in every particular, proving that the addition of minerals and nitrogen is absolutely essential.

The necessity of each of the mineral constituents, and

their influence, are shown by omitting in each case, in another series of experiments, one of the mineral elements; the crop secured in each experiment is found to be imperfect. Perfect growth is only reached when all the elements named are present.

Food obtained from the Atmosphere.— The leaves and roots are called organs of nutrition; the leaves take material from the air, the roots from the soil.

The dry substance of a plant is made up chiefly of carbon, and the proper absorption of this element depends upon the activity of the leaves. The leaf is made up of rows of cells, placed side by side, which in the under portion are loosely held together, forming "air spaces" between them, and over the whole leaf there is a thin, transparent skin. This skin, called the "epidermis," is not entire, but is dotted with little holes called "stomata;" through these the atmospheric air passes freely into the cellular spaces in the leaf, and through the porous walls of the leaf cells, which contain a green substance called "chlorophyll."

In these cells, containing chlorophyll, the carbonic acid of the air is broken up through the influence of light and warmth, and the oxygen set free and the carbon retained. This process is called "assimilation," and is active only during the daytime; during the night the food, formed by day, undergoes a change, whereby it can be transferred in solution to the places where it is needed. Daylight, as also the presence of iron in the plant, is necessary for the formation of chlorophyll.

In the daytime, growing plants purify the air by consuming the carbon; in the night, the process is reversed, oxygen being then consumed. The quantity of oxygen set free by young plants is, however, greatly in excess of that consumed. The food directly supplied by the atmosphere is, therefore, chiefly carbon.

The carbon absorbed by the plant, in connection with the hydrogen and oxygen of the water, forms starch, dextrine, sugar, fat, cellulose, substances serviceable in the growing plant, and to be stored away for future use in seeds, roots, and stems; and the same elements, together with nitrogen and a little sulphur, form the albuminoids. All of these are vegetable substances with which we are familiar, and which are termed "organic."

Food obtained from the Soil. — Roots are of two kinds, — the "tap root," the chief use of which is to sustain the plant in an upright position; and the "fibrous root," which is engaged mainly in absorbing food from the soil. These roots are made up of cells, through the walls of which solid matter cannot pass; all food absorbed by the root must be in solution. The surface membrane of the root, unlike that of the leaf, is not full of holes; the absorption of solids is, therefore, impossible. The absorption of the solution by the root is obtained by means of what is called "diffusion." By this means water is absorbed, and, with the water, the dissolved food elements usually contained in the ash of plants. The absorption of food is, however, not confined to the taking up of ready formed solutions.

The root grows at its tip, and it is through the delicate cells located there that the plant absorbs nutriment. The tips are protected by a sheath, or cover, of dead or dying cells, which protects the delicate fibres, and permits the root to penetrate the soil without injury. The acid sap or fluid which is produced by the root, hair, or cell, when it comes in contact with soil particles, has a solvent effect upon them; thus is insoluble matter in the soil, containing ash ingredients, made soluble to the plant.

These elements derived from the soil are, however, not taken up in the form of individual chemical elements, but chiefly as acid or salts. Nitrogen, for instance, is combined with oxygen to form nitric acid; which, united with bases like sodium or calcium, forms sodium or calcium nitrates. The very weak solutions of the compounds taken up by the roots are concentrated in the upper part of the plant, owing to the rapid evaporation of the water through the leaves, and are employed in the formation of new tissues.

Supply of Food Constituents. — The atmosphere contains relatively a very small proportion of carbonic acid, from which the carbon is obtained; it is less in the open country and over large bodies of water, and more in the vicinity of towns, yet the supply in the aggregate is enormous, and is sufficient to meet all the demands of vegetation for carbon.

The atmosphere also contains small quantities of nitrogen, existing both as ammonia and as nitrates; these are brought to the earth by means of dew and rain, and thus act as a direct source of this element to plants. The amount so provided is, however, insufficient for the entire needs of vegetation.

The moisture, or "water vapor" in the atmosphere does not, to any great extent, serve to supply the plant with water, the absorption of water being a function of the roots rather than the leaves. The leaves absorb water when there is an insufficient supply in the soil, or when the weather is such as to cause a very rapid exhalation from the leaves. The supply of the mineral constituents depends upon the character and composition of the soil, upon the season, and climate; these conditions are not uniform, hence the increased variation in the character of the natural vegetation found in different parts of the world.

Functions of the Constituents.— The exact work that each constituent performs in plant nutrition has not yet been definitely determined. It has been shown that nitrogen is of vital importance, since it is an essential constituent of the living principle of plants called "protoplasm." Phosphorus and sulphur have been found essential constituents in the formation of albuminoids, a very important compound of all plants. Potash is necessary to the formation and distribution of starch, while the green color of plants, or chlorophyll, cannot be formed without iron. All the constituents mentioned are necessary, and are concerned in the various processes which result in the perfect plant.

The Natural Tendency of Plant Growth.— The ultimate object of all plants in their natural state is to form seed, or the beginning of new plants. The seed of every plant contains within itself a sufficiency of food to nourish the germ till the root and leaves of the new plant are sufficiently developed to acquire food from the sources mentioned.

Germination : the Conditions Necessary.— The first evidences of growth are shown in the germination or sprouting of the seed. This beginning depends chiefly upon three conditions, — first, the presence of moisture; second, of warmth, or a proper degree of temperature; and third, the access of air. The seed of the wheat plant, for instance, will retain its form and remain a seed for a long time if kept dry. As soon as it is enabled to absorb moisture it increases in size, oxygen is absorbed from the air, and heat develops; it separates the germ, and is no longer a seed, but a young plant. Seeds do not germinate below a certain temperature, usually about 37° F.; the warmth necessary for germination is, however, different for different seeds. Wheat will not germinate below 41° F., and corn below 49° F. There is also a certain temperature above which seed will not germinate; the higher limit is 108° F. for wheat, and 115° F. for corn. The temperature of most rapid germination lies between 79° F. and 94° F.

The presence of air is universal, and care in this respect is only important when plants are grown in comparative confinement.

Duration of Life. — Plants which germinate, grow, and produce flower, fruit, and seed in one year, and then die, are called "annuals." Wheat, rye, oats, buckwheat, peas, and beans are good examples of annual plants. Certain other plants require two seasons for this work; these are called "biennials." During the first season, the organs of growth are developed, viz., the root and leaf; the second season, the flower, fruit, and seed are formed, after which the plants die. Examples of biennial plants are cabbages, turnips, parsnips, celery, lettuce, beets, etc.

In order to secure seed it is not necessary that the plants remain in the ground throughout the winter. Certain quick growing biennials, as radishes, may produce seed in one year, if removed from the soil after the root is full grown, topped, and transplanted.

"Perennials" are plants that live for more than two years; timber-trees, fruit-trees, berries, grape-vines, etc., are good examples, though perennials are not confined to trees and shrubs; asparagus, a number of the clovers, and various grasses, also belong to this group.

Plant Development. — The development of the plant after germination is not uniform; the substances obtained by the roots are greatest in the young plant, which is always rich in nitrogen and ash elements. As the plant grows, the proportion of food derived from the atmosphere through the action of the leaves steadily increases.

The wheat crop contains practically all of its nitrogen and potash when in full bloom; carbon increases as long as the plant remains green. When the seeds begin to form, the food gathered by the leaf and root is largely transferred from the stem and leaf of the plant, and concentrated in the seed.

Cereals and grasses cut while the crop is green are much richer in nutritive matter than when they are allowed to ripen seed. In such crops as turnips, beets, and potatoes, the development of root and leaf is the same as in wheat; but at the completion of growth, food is stored up or contained in the root, or tuber, and the leaves die after the food in the plant has been largely transferred. In trees, the plant-food gathered by the root is concentrated by the end of summer in the pith of the tender branches and in buds, and serves as food for the new growths of another season.

CHAPTER II.

Origin and Formation of Soils.

Soils: Their Origin, Formation, and Classification.
— Every growing thing can be traced back to two primary sources, — the atmosphere and the soil. Every chemical element contained in plants or in animals produced from plants can be found either in the soil or atmosphere.

The constituents which plants derive from the atmosphere are so abundant everywhere that the continuous growth of maximum crops cannot exhaust them, and no particular efforts are required to increase their efficiency to the plant. The constituents of the soil are much less abundant, and the power of the plant to secure them depends very largely upon the effort of the farmer. To the farmer, then, the soil is the object of first attention.

What is a Soil? — The soil is the name given to that part of the earth that can be cultivated, and in which plants can grow.

Origin of Soils. — Soils are derived, directly or indirectly, from the rotting or decay of rocks. If all the earthy matter which composes soils could be removed from the surface of the earth, the remainder would be solid rock. Rocks have been formed by the action of fire and water; hence they are classed as "igneous," that is, produced by the action of fire, or "aqueous," produced by the action of water. The igneous rocks form but a small proportion of the outer rocks of the world; yet they are of great importance, because it is through their decay or breaking up that the other rocks have been formed.

The most important of the igneous rocks are granite and trap. These contain silica and alumina as their chief constituents, and greater or less amounts of potash, lime, magnesia, iron, and the other mineral constituents of plants. These rocks are also called "primary rocks," because they are supposed to constitute the primary or first crust of the earth, and to be similar in character and composition to the now molten interior of the earth. This view is based upon the fact that the lava ejected from active volcanoes resembles in composition the trap rocks.

The trap rocks consist mainly of two minerals, feldspar and hornblende. Feldspar is particularly rich in potash and soda, and poor in lime and magnesia. Hornblende is poor in potash and soda, and rich in lime and magnesia. Granite rocks consist mainly of quartz, feldspar, and mica. Quartz is almost pure silica; while mica contains nearly all the compounds found in both feldspar and hornblende, and is one of the most abundant minerals.

The aqueous rocks are formed from the minerals contained in igneous rocks, the proportion of the different constituents contained in them depending upon the method of their formation. Limestone and red sandstone are good examples of aqueous rocks.

Subsoil. — The subsoil lies immediately underneath the soil, and rests upon the solid rock. The main distinction between soil and subsoil is that the soil contains more organic or vegetable matter, is more finely divided, and is less compact than the subsoil. The subsoil may be regarded as something between the soil and rock, and partaking in part of the characteristics of both. The subsoil serves to gradually supply the constituents that are removed by crops from the surface soil, and also performs important functions as a reservoir of moisture, and as a medium for the roots of plants. Its character thus materially modifies the fertility and productiveness of the soil proper.

The Formation of Soil. — Certain agencies are constantly at work converting rock into soil. This gradual conversion is termed "weathering," and is observed on all sides. The rapid crumbling of shale, so familiar in New Jersey, on exposure to the atmosphere, is an excellent illustration of this process, — the air, temperature, and water all playing an important part.

The Action of the Atmosphere. — The atmosphere causes what is termed "oxidation," a slow burning or decay. The oxygen of the air combines with the minerals, forming new substances called "oxides." A familiar illustration of this process is seen when iron is exposed to the air; the red rust that forms is an oxide of iron, a substance very different from the iron itself.

The Influence of Water. — The most powerful agent in the rotting or breaking up of rocks is water. It acts both mechanically and chemically. Most rocks contain cracks or fissures into which the water readily penetrates; by freezing the water expands, and forces the rock apart. Thus the solid rock is gradually separated into fragments of different sizes. The force of the water in rains and streams grinds the fragments together continually, wearing them smaller, and the smallest are first carried away by the water to lower levels.

The overflow of rivers always leaves a deposit of soil particles, more or less finely divided, carried in suspension in the water; this sediment gradually accumulating forms soil. What are termed by geologists "sedimentary, or aqueous, rocks" have also been formed by the action of water, followed by heat or pressure, or both.

Water also has a decided effect chemically. The carbonic acid in the water absorbed from the atmosphere dissolves certain of the chemical ingredients of the soil, particularly lime, though its solvent effect is not confined to this substance; soda, potash, silica, and iron are also attacked to a greater or less degree.

The Growth of Plants. — The growth of plants is also active in the rotting of rocks. In their growth the roots penetrate the crevices and force the particles of rock to separate; they also attack and absorb certain portions of the constituents that are made soluble. In the decay of plants, the soil is kept moist, gases are generated and absorbed by the water, which, again penetrating the rocks, has a solvent effect upon them.

It is a matter of common observation to see rocks covered with the lower orders of plants, like lichens and mosses; the removal of these frequently shows the rocks furrowed and eaten into by the roots, due to the action described. This growth and this decay of vegetable life, continued through a long series of years, gradually deepen the soil, and prepare it to produce the higher order of agricultural plants, or those useful as food for man.

Earth Worms.— Earth worms and other living creatures living in the ground also aid in preparing the soil for the growth of plants. They burrow deep into the ground; the passages thus made aid in drainage and circulation of air, and facilitate the penetration of roots, which increases their power to secure food. They also drag into their underground passages considerable vegetable matter, which in its decay aids in forming true soil.

Movement of Soils.— While our present soil remains almost constant in quantity, the parts removed by various causes being supplied by the gradual decay of the rocks, the same agencies which form soils, particularly weather action, are wasting or carrying them away. Soil is almost constantly moving; it is thinnest at the top of the hill, and deepest in the valley. This is very noticeable in mountainous or hilly districts; great furrows are formed in the hillsides after heavy rains, the swollen and muddy streams carrying the soil in suspension to lower parts of the land.

Soils in Place.— Soils which have been formed from the decay of the underlying rock are called sedentary soils, or soils in place; they have not been moved from the place in which they originated. These partake of the nature and composition of the rock underneath; though, from the method of their formation, viz., the growth and decay of plants, they contain considerable vegetable matter obtained from the atmosphere.

Transported Soils. — Transported soils are those which have been moved from the place where they originated, and deposited from water or ice. The former are called "alluvial;" these occur not only in valleys and river-beds, but in inland places, where they have been deposited in lakes or other bodies of water. Those deposited from ice are called "drift soils;" these have been formed by the action of "glaciers," vast bodies of ice moving like a river, which carry considerable quantities of earth as well as numerous rocks.

The results of glacial action are found, not only in valleys, but in high mountains, where rocks show, by well-defined scratches, the wearing and grinding effect of the moving ice.

Drift soils are distinguished from all others by the presence of rounded rocks or bowlders, and, though not uniform either in composition or character, are usually fertile, the fertility being due chiefly to the bringing together, from numerous sources, of a variety of mineral substances.

Classification of Soils. — The principal ingredients of soils are sand, clay, carbonate of lime, and vegetable or organic matter. They are, therefore, divided into four natural classes; viz., sandy, clayey, limy, and peaty, according to the greater proportion in each case of one of these four ingredients.

What is Sand? — Sand or silica is composed of the mineral silicon united with the chemical element oxygen, and does not serve directly as food for plants. A soil consisting entirely of sand would be useless to the farmer, for he could produce nothing from it in its natural condition. It possesses distinct properties, however, which render soils containing considerable sand light and open, and therefore permeable to air, moisture, and warmth. The stony particles of sand are also rapidly heated by the rays of the sun, thus very greatly influencing the temperature of the soil.

What is Clay? — Pure clay contains silica, together with alumina, a compound of aluminum and oxygen. From a chemical point of view, pure clay would be quite as useless as sand as a source of plant-food, neither silica nor alumina being essential constituents of plants. The properties of clay are almost the reverse of those of sand. Particles of sand do not adhere to each other — moist sand pressed firmly in the hand will fall apart immediately the pressure ceases; particles of clay, on the contrary, readily adhere to each other, and, when moist, can be moulded into any desired shape, which is retained; on heating, the particles adhere still more strongly, a characteristic taken advantage of by the manufacturers of brick, tile, etc. Sand from its porous nature rapidly loses water; clay from its compact nature retains moisture. Sand absorbs heat rapidly, and soon becomes dry; clay absorbs heat more slowly, and remains cool. Clay, like sand, is, however, a valuable constituent of soils; its tenacious character enabling it to retain both moisture and the useful constituents applied in manures.

Lime. — The lime of soils usually exists in the form of carbonate of lime, or limestone. Limestone is a valuable ingredient of soils, not only because it furnishes the important constituent of plants, calcium, but because of its relative ease of decomposition, and of its valuable action upon and reaction with other soil constituents; it aids in the decay of vegetable matter, and in the formation of nitrates. It exerts a favorable physical effect upon soils; its presence helps to separate the adhesive particles of clay, and makes heavy soils loose and friable, which permits the easy passage of water through them. Lime also increases the absorbing and retaining power of sandy soils, by causing the particles to adhere more closely to each other.

Humus. — The decaying vegetable matter in soils, which is made up of carbon, oxygen, hydrogen, and nitrogen, is called "humus." In virgin soils it is derived from the dead roots and leaves of a former vegetation. It has a dark brown or blackish color. Leaf mould, found in forests, is largely composed of humus. It was believed at one time that humus served as a direct food for plants, but this idea has been proved to be incorrect; it is the final products of its decay, chiefly carbonic acid, ammonia, and water, that serve as plant-food.

Humus is a very useful ingredient in all kinds of soils, though a soil may contain too much decaying organic matter. Humus improves sandy soils, not only on account of the nitrogen and other plant-food constituents which it contains, but because it increases their absorbing and retaining power. Humus will absorb and retain more moisture than any other ingredient of soils. Clay soils are improved by it, however, on account of its property of loosening and aerating them.

The four principal ingredients of soils are useful, therefore, not altogether because they furnish plant-food, but because they give to soils certain physical properties which enable them to retain heat, moisture, and plant-food. These properties give to soils what is called physical character, which is very important in determining fertility.

Sandy Soils.— Land which contains over seventy per cent of sand is called sandy. Such soils are not only poor in plant-food, but they can absorb and retain but little moisture. The soil, loosely held together, permits the rapid passage of water, and the stony particles readily absorb heat. In hot, dry seasons, the crops are soon parched; in wet seasons, however, these same properties enable the soil to dry quickly, thus permitting, if sufficient food is provided, the growth of maximum crops, when they would be destroyed from excess of moisture on soils of a more dense or tenacious character. Sandy soils are easy to work, and may be cultivated when quite wet without injury, and are well adapted to quick-growing crops; when overlying clay subsoils, they are susceptible of a high degree of fertility.

Clay Soils.— A clay soil is one which contains over fifty per cent of clay. A clay soil is almost the reverse of a sandy soil. The finely divided particles adhere so closely as to make the access of air, moisture, and warmth, difficult; they are, therefore, called cold and tenacious. They are hard to work, and, unless well drained, crops are liable to suffer both in wet and dry seasons; in wet seasons because the moisture is not freely movable, and in dry seasons because the land becomes so hard as to prevent the penetration of roots. They are well adapted to the growth of cereals and grasses.

Limestone Soils.— The term "lime" or "calcareous," is applied to a soil when it contains over twenty per cent of lime. Limy soils are usually of a good character, are easy to work, and well adapted to cereals or fruit.

Peaty or Vegetable Soils.—A peaty or vegetable soil consists almost entirely of vegetable matter, more or less decayed. Such soils are very productive if well drained, and furnished with a sufficiency of the mineral constituents.

Soils, however, which are best adapted to the purpose of farming do not belong strictly to either of these classes, but are rather those which contain more even proportions of the ingredients; hence, a further division of soils is usually made, based upon the relative proportions of the principal ingredients, sand and clay.

Loamy Soils.—A soil consisting of a mixture of sand and clay is called a loam; if it contains from ten to twenty per cent of clay, it is a sandy loam; if from twenty to thirty per cent of clay, it is a loam; if from thirty to fifty per cent of clay, a clay loam. Gravelly or limy loams are those in which gravel or coarse sand and lime are contained in considerable amounts. Loams are suitable for most purposes, and soils of the different natural classes are improved as they approach the character of loams.

It would be impossible to describe and classify the almost infinite variety of soils which exist in nature, and which are made up from different proportions of the four ingredients mentioned.

Perfect Soil.—A perfect soil is one which contains the ingredients in perfect proportions: sand, to enable it to absorb air and moisture in proper amounts, and to render it warm and friable; clay, to keep it cool, and prevent a too rapid leaching or evaporation of water; lime, to assist in the decay of vegetable matter; and humus, to retain the best amount of moisture, and to furnish material for the various chemical processes continually going on in good soils.

These conditions are rare, and seldom occur in nature, though it is in the power of man to produce them: still, perfect soils are not to be had without a great deal of labor and care; and frequently it is more important for the farmer to adapt himself to his soil, and make it produce the best which, from its natural character, it is best capable of doing, than to attempt to change its character. He should not attempt to raise wheat when the soil is peculiarly fitted to grow early vegetables or fruits, nor to grow early vegetables on soils only adapted for the grasses.

CHAPTER III.

Composition of Soils.

A SOIL, like a plant, consists of two distinct classes of substances, — first, organic or vegetable, derived, as we have seen, from decaying growths; second, inorganic or mineral constituents, derived from the rocks which form the earth's surface.

Organic Substances. — Organic substances are made up of carbon, oxygen, hydrogen, and nitrogen. Of these the nitrogen is of the most direct importance in the growth of the plant, and is the valuable constituent of the humus already described.

Inorganic Substances. — The inorganic or mineral substances of the soil are also identical with the substance of the ash of plants (with the addition of alumina, which is not taken up by the latter), namely, silica, alumina, lime, potash, magnesia, phosphoric acid, soda, iron, chlorine, and sulphuric acid. The first three of these, as has already been noted, are the principal ingredients of soils, and give to them their distinctive character; with the exception of lime, they do not aid materially in furnishing food. The more important constituents are phosphoric acid and potash.

Phosphoric Acid. — Phosphoric acid is an ingredient of all fertile soils, but is contained in very small quantities as compared with other constituents. Its most common combination is with lime, though it is frequently found in combination with iron and aluminum. Rocks which contain "fossils," or fossiliferous rocks, frequently contain high percentages of phosphoric acid.

Potash. — Potash is also derived from rocks, and varies in the amount contained in different soils. Those derived directly from granite or trap are the richest in this element; it exists in the soil in combination with silica, forming substances called "silicates," which are of great importance.

Lime. — Lime is an ingredient of most soils, and is derived from the decay of limestone, or from fossils.

The Natural Fertility of Soils. — The mineral constituents, phosphoric acid and potash, though contained in soils in relatively small amounts, ranging from less than one-tenth per cent to over one per cent, give to soils their chief claims to natural fertility; since most agricultural plants require relatively large proportions of these in proportion to other mineral constituents.

The quantity of phosphoric acid and potash contained in a soil is, however, comparatively great, since the surface soil at a depth of nine inches will weigh, when perfectly dry, three to three and one-half million pounds per acre; hence, with even one-tenth per cent, it would contain from three thousand to three thousand five hundred pounds of each of these constituents.

The other necessary mineral ingredients are found in greater or less, and usually in sufficient amounts in all soils. These, while all essential to the complete development of the plant, are of course less liable to exhaustion.

The immediate fertility of a soil depends, however, not so much upon the quantity of the constituents contained in it as upon the amount of each that may be available to the plant.

Analysis of Soils. — The composition of a rich wheat soil and of a wheat plant, as shown by analysis, in the diagrams[1] (page 32), indicates the relation of the composition of the plant to the composition of the soil.

The chemical analysis of a soil shows the percentages of the different constituents contained in it. It is observed that the constituents which the soil possesses to only a limited extent are contained in the plant in relatively large amounts; these are, therefore, termed essential plant-food constituents, because of their greater liability to exhaustion.

Weight of Soils. — In studying a soil from a statement of its analysis, regard must be had to the weight of soils. The constituents of the analysis are expressed in per cent or pounds per hundred; it is evident, therefore, that a soil weighing one hundred pounds per cubic foot, and containing four-tenths of a per cent of phosphoric acid, would contain a much greater amount of phosphoric acid per acre than a soil showing the same percentage, but weighing fifty pounds per cubic foot.

It is estimated that dry sand weighs from one hundred to one hundred and twenty pounds per cubic foot; loam, from ninety to one hundred pounds; clay, seventy to eighty pounds; and peat, thirty to fifty pounds. An analysis, therefore, which shows the same per cent of the constituents in one soil may not indicate its

[1] Adapted from Ville.

Composition of Rich Wheat Soil.

POUNDS PER HUNDRED.

Carbon.
Hydrogen . . . 12.67 } Elements which, though essential, are abundantly supplied by the air and water.
Oxygen.

Silica 71.55
Alumina 6.94
Iron . . . . . 5.17 } Elements which are either non-essential or are required by the plant in minimum amounts.
Magnesia . . . 1.08
Soda 0.43
Sulphuric Acid . 0.04
85.21

Nitrogen 0.12
Phosphoric Acid . 0.43 } Elements absolutely essential, of which relatively large amounts are required.
Potash 0.35
Lime 1.22
2.12

Composition of Wheat Plant.

POUNDS PER HUNDRED.

Carbon 47.69
Hydrogen . . . 5.54 } These are derived from the air and rain.
Oxygen 40.32
93.55

Soda 0.09
Magnesia . . . 0.20
Sulphuric Acid . 0.31 } These are abundantly provided by the soil, and it is unnecessary to add them in any case.
Chlorine 0.04
Iron 0.06
Silica 2.75
3.45

Nitrogen 1.60
Phosphoric Acid . 0.45 } These the soil possesses only to a limited extent, and the deficiency must be supplied.
Potash 0.66
Lime 0.29
3.00

true character in reference to the amount of plant-food contained in it, unless it is accompanied by a description of its general character, whether sandy, clayey, or peaty.

It will be observed from the foregoing, too, that the ordinary idea of the weight of soil has reference to the physical character, rather than actual weight. A sandy soil is called "light," and a clay soil "heavy;" while in reality a sandy soil is heavy, and a clay soil is light.

The sandy soil is called light because it is easier to work, — the particles of sand are readily separated from each other, and no particular force is required; while in the case of clay soils the particles adhere tenaciously, and it requires considerable force to separate them.

Classes of Soil Constituents. — The constituents of a soil may again be divided into three classes, — first, those which serve mainly as a mechanical support for the plants, like sand, clay, limestone, and gravel; second, dormant or reserve substances, which not only act the same as the first class, but are capable of changing to such a form as to furnish nutrition to the plant; these are of both vegetable and mineral origin; and third, active constituents, or those directly available to plants.

The mechanical constituents constitute the bulk of all soils — frequently over ninety-five per cent. The second class, or dormant, are contained in most soils in relatively small amounts as compared with the first class, though the quantity that a soil may contain varies considerably, depending upon its formation.

The active constituents, those that are immediately available to the plant, are never present in even the best soils in large amounts; they are formed slowly from those that are dormant. Humus is not a direct food, but is capable of being changed into food.

Clay and substances containing phosphoric acid, by weathering, or the action of frost, heat, and moisture, are changed to such an extent as to give up, in time, portions of their potash, phosphoric acid, and lime.

The True Measure of Fertility. — The active constituents, however, measure the true fertility of any soil. The dormant substances may be rich in phosphates, potash, lime, and humus, and yet it may be impossible to produce a single plant from them, because the surrounding conditions are never favorable for the activities that cause their change into active substances.

An analysis of the soil does not show true fertility, unless it shows how much of the total constituents of the soil are capable of being made active, and thus useful to the crops: it simply shows the possibilities that are lying dormant.

One Element Cannot Substitute Another. — Another point is also important in this connection; namely, that, of the three active constituents, nitrogen, phosphoric acid, and potash, which exist in small quantities in all soils, the one contained in minimum amounts in the soil determines its power of producing plants; that is, the crop cannot rise above the point measured by the element existing in the smallest amount — one element cannot be substituted for another. For example, if we have in an acre of soil only sufficient nitrogen for ten bushels of wheat, the crop could not be increased to any considerable extent beyond that point, even though phosphoric acid and potash were contained in unlimited quantities; a balance of the plant-food constituents is essential to full and complete growth and development.

Exhaustion of Soils. — Exhaustion of soils has reference mainly to the four constituents, nitrogen, phosphoric acid, potash, and lime; the amount of the others is usually contained in excessive quantities in all soils. Exhaustion is, however, a relative matter, since it is not possible to completely exhaust a soil of its active constituents. Exhaustion means properly the reducing of the constituents to that point which makes the production of crops unprofitable; hence the question of exhaustion is a variable one, determined in a large measure by local circumstances.

Exhaustion, too, may have reference to one constituent only; for instance, there may be an abundance of nitrogen and phosphoric acid, and a deficiency of potash. By growing a class of crops which take more of the constituents that are present in relatively large amounts, and less of those that exist in small amounts, the period of exhaustion is deferred.

Natural Strength of Soils. — The power which soils possess of gradually forming active ingredients is termed natural strength. It is obvious that the character and origin of the soil have an important bearing upon this point.

The natural strength of a light sandy soil may be measured by a crop of wheat of five bushels per acre; while the natural strength of rich valley or prairie soil may be measured by an annual yield of twenty-five bushels of wheat per acre; that is, in the one case, the substances which form the soil are of such a character as to permit of the change of but a small proportion of its dormant into active constituents, while, in the other, a large proportion of the constituents are annually rendered available.

Soils overlying limestone and granite, and those formed by the gradual accumulation of vegetable matter, as in the prairies of the Western States, possess a high natural strength. They contain large quantities of the dormant constituents, which are of such a character as to be readily changed into activity under ordinary conditions of season and farm practice, and large crops are possible each year for a long period.

Sandy soils, perhaps, are the best examples of soils of low natural strength; in these the purely mechanical constituents are in great excess, no considerable quantity of dormant or reserve substances exist, and the constituents made active are only sufficient for minimum annual yields.

Texture of Soils. — Another point to be taken into consideration, when studying the composition of a soil, is the power it possesses of absorbing and retaining such constituents as may become active. This characteristic of a soil is termed its "texture;" and it has not only a bearing upon the adaptability of the soil to the growth of plants, but also exercises a decided influence upon such growth.

Soil so open in texture as to freely admit the circulation of water is more liable to be depleted in its active constituents than a soil which is close in texture, and retains for a considerable time the water which falls upon it.

The nature of the subsoil is also an important consideration. If the soil rests immediately upon a rock, or upon sand, it will be found to dry out much more rapidly than if it rests upon a clay subsoil. In the first place, the water passes rapidly beyond the reach of the roots, and cannot readily get back; in the second place, it does not percolate so rapidly, while at the same time it retains its connection with the surface.

This point is very apparent to one who has observed the red shale soils in central New Jersey. They are reasonably rich in all forms of plant-food, yet those which lie directly upon the shale or rock are much less productive than those which lie upon a clayey subsoil. The shale permits a too free escape of water, and crops suffer more severely from drouth than those which lie upon a clayey subsoil. It is essential that water be freely movable in soils, in order to properly prepare the food, as well as to carry it to the roots of plants, but it must be freely movable in all directions.

Climate. — The climate is also a matter of importance: rainfall, temperature, location, all exert an influence in determining the value of a soil, and should be taken into consideration in connection with its chemical composition.

The average rainfall may be sufficient; but if it is not properly distributed throughout the growing season, normal growth is impossible. The same is true of temperature; cold in harvest time is ruinous, though the normal temperature for the year may have been attained.

Loss and Gain to Soil. — A soil, whether cultivated or not, is continually changing, the various causes which combine to form soils being ever at work to make them richer or poorer. If left to themselves, the constituents rendered soluble by air, moisture, and chemical action, as well as the finer particles of earth, are carried by rains in greater or less amounts into the streams and brooks. Certain of the soil constituents are, however, less liable to be lost through drainage than others; that is, soils do not exert the same retentive power for all constituents.

The constituents of the greatest interest to the farmer are nitrogen, phosphoric acid, potash, and lime. Of these, nitrogen and lime form certain compounds that are extremely soluble and freely movable in the soil; drainage waters are seldom free from traces of nitrates, and of chlorides of sodium and calcium (lime).

On the other hand, ammonia, a compound containing nitrogen, and phosphoric acid and potash are seldom found under natural conditions in any considerable amounts in drainage waters. For these the soil possesses a strong retentive power, though they are not held so strongly as to be unavailable to plants.

This power of soils is not only important in showing the probable loss or gain of fertility in uncultivated soils, but has a wide bearing upon their possible improvement. Phosphoric acid and potash particularly, when added to soils, are fixed, and remain until removed by the plants.

Clay, humus, and lime are the ingredients in soils which exert the greatest influence in retaining the soluble phosphates, potash salts, and ammonia compounds.

Absorptive Properties of Soils. — The property which a soil possesses of breaking up such compounds, and holding fast to the essential elements, is both physical and chemical. The holding of such bases as potash, lime, etc., is due to the presence in the soil of what are termed simple silicates; these are capable of combining with other silicates to form double silicates. A silicate of alumina, for instance, will combine with a silicate of ammonia to form a double silicate of alumina and ammonia. All soils possess this absorbing power in some degree, though it belongs particularly to soils containing clay.

Soils do not, however, possess an equal absorbing power for acids. Nitric acid is not absorbed, but is freely movable. The only acid of importance absorbed by the soil is phosphoric acid, which combines with lime, iron, and alumina, forming phosphates, — compounds of great importance in plant nutrition, which are not removed from the soil except through the growth of plants.

This absorbing property of soils may be nicely illustrated by filling a cylinder of suitable length with a good soil, and pouring upon it a dilute solution containing one or more bases, including potash and lime, and both nitric and phosphoric acid. An examination of the solution which passes through will show the presence of nitric acid, and an absence, at least in any amount, of the potash and phosphoric acid. Good soils fix all of the essential constituents, except nitrogen when it is in the form of a nitrate.

The farmer can reduce the losses due to drainage by careful management. The drainage waters contain least nitrates when crops are growing and well cultivated. This carrying away of plant-food constituents by the rain into the drains may be regarded, therefore, as a natural loss to soils, and is greater or less according to the character of the soil, and the treatment it receives. On the other hand, there is a gain in the fertility of soils due to natural causes.

Rain carries to the soil appreciable amounts of nitric acid and ammonia, as well as certain solid substances, existing in the atmosphere. The gain from this source is greatest in the vicinity of cities, and least in the open country. The gain due to the action of water, heat, cold, and decaying vegetable matter has already been referred to in previous sections, though the changes taking place in vegetable matter require further notice.

Nitrification. — Vegetable matter is the source of humus of soils, and the active principle of humus is nitrogen. The nitrogen in humus is combined with carbon, and in this form it is not available to plants. In order to become most useful to them it must be changed into a nitrate, since plants take up their nitrogen chiefly in this form. This process is called "nitrification," and is caused by minute "organisms" or "ferments," which are present in all fertile soils.

These ferments are most active in warm, moist, well-drained soils, when nitrification proceeds rapidly; they are not active when the temperature is lower than $41°$ F. or higher than $131°$ F. In winter, in temperate climates, nitrification practically ceases altogether, while in summer it proceeds most rapidly. As soon as nitric acid is formed by this process, it immediately combines with some base, preferably lime; hence, if drainage is allowed, the loss of nitrates is always accompanied by a loss of lime.

CHAPTER IV.

The Improvement of Soils.

The improvement of soils may be regarded as of two kinds, — first physical, and second chemical; though this classification is not always well defined. Frequently an improvement in the physical character of soils is also accompanied by important chemical changes. The improvement of soils due to natural causes, while considerable in the aggregate, is insignificant, in point of time, compared with that which may be secured by the farmer through artificial means. The true aim of the farmer should be to bring the soil into a condition to produce crops which are well adapted to his location, and which are as large as the average conditions of climate and season will permit.

The first point to determine is whether the land is worth improving; the kind of crops that can be raised, and their probable market value, must guide in this respect.

Physical Imperfection. — One of the chief imperfections in natural soils, aside from their chemical character, is in respect to water; they contain too much or too little. If too much, the imperfections may be in many cases corrected by proper drainage; if too little, by adding water or such materials as may increase the absorbing and retaining power of soil for water.

The earth may be compared to a sponge full of water, which rises towards the surface with heavy rainfalls, and falls below as evaporation and percolation proceed.

Drainage. — If a hole dug into the soil partly fills with water, and remains with slight fluctuations through the season, the water contained in it is called "bottom water," and the point to which it rises is called the "water level." If the water level is constantly near the surface, the soil is liable to be too wet; for most plants suffer if their roots are immersed for any length of time in stagnant water. Plants need air, both for root and branch. Too much water in a soil prevents the circulation of air, and also keeps it too cold for most crops; the cranberry and rice plants are prominent exceptions to this rule. The soil may also be too wet, even when the water level is deep into the earth, by reason of absorbing too much of the rain that falls upon it. Drainage corrects in the first case by lowering the water level, and in the second by permitting a more rapid passage of water through the soil.

Land well drained is improved, not only by the removal of water from it, but because the more rapid diffusion and passage of the water through the soil carry the air and warmth to lower levels, which are, as has already been shown, important factors in making soil constituents soluble, and thus increasing the power of plants to secure food.

In too many cases half-developed crops are secured year after year upon land, which, if properly drained, would be capable of maximum production. Where springs occur, and where the land is composed of clay overlying clay subsoils, drainage usually results in great improvement and profit to the owner.

Methods of Drainage. — The efficiency of drains depends upon the free passage of water through them. They should always lead to the lowest portion of the field; if the land is level, they must be gradually sloped — one foot in five hundred will furnish sufficient grade for the flow of water. On a slope, the drains may be laid at a uniform depth from the surface; the main drain should always occupy the lowest part of the field.

The depth of drains and their distance from each other are governed by the character of the land. On light, open soils, they should be deeper and farther apart; on heavy land they should be nearer to the surface and to one another. The mouth of the drain should be well protected, and kept free from all obstructions.

Irrigation. — When lands contain too little water irrigation is frequently resorted to, though the best results from irrigation are attained on well-drained land. Irrigation not only softens the land, thus making it more permeable for the roots of plants, but it is effective in dissolving the dormant constituents of soils. Large tracts of now barren land in the United States only require water to make them fruitful.

The advantages of irrigation are, perhaps, most conspicuous in the States of Colorado and California. In portions of the Eastern States crops are frequently ruined by a lack of water at the right time; the irrigation of these areas is only a question of time.

Claying and Sanding. — Further imperfections in the physical character of natural soils are also common.

Sandy soils are improved when made more compact and tenacious; this may be accomplished by adding clay, or organic matter, or both. Clay soils are improved as they are made more porous and open; this may be accomplished, in part at least, by the addition of sand. Claying and sanding are expensive processes, and are seldom resorted to in this country except in cranberry culture; though in districts where clay marls are abundant, the same object is accomplished by the application of this material. Marling, however, materially improves their chemical character, because of the mineral constituents, potash, phosphoric acid, and lime, one or all of which may be contained in them.

Green Manuring. — The addition of organic vegetable matter to soils, for the purpose of improving both their physical and chemical character, is readily accomplished by means of green manuring. The term "green manuring" is used when the crops themselves are plowed under in their green state. Any plant, of course, may serve for this purpose, though those most commonly used are red and crimson clover, cow pea, rye, and buckwheat.

Plants most Useful. — Of these crops the clovers and peas are more useful than the others. Clovers, peas, beans, lupins, vetches, and a number of others of less importance, belong to a class of plants called "legumes," which have the power of securing nitrogen from the air, and can, therefore, make perfect growth under proper conditions without depending upon soil nitrogen. This function of the legumes has long been known by practical farmers, but the method by which the nitrogen is obtained is a quite recent discovery.

These plants have small, knotty growths, called "tubercles," on their roots. which are believed to be caused directly or indirectly by certain bacteria which are present in soils in which this class of plants are grown. Recent experiments have shown, too, that soils which do not contain these bacteria may be inoculated by applying a light dressing of soil from a field in which the plants have previously grown to perfection, without direct applications of nitrogenous material. It is through these tubercles that the plants are supposed to gain free nitrogen from the atmosphere. Experiments have shown pretty clearly that, where they have been formed, the nitrogen in the crop is far greater than when they are absent. The fact that such is the case is sufficient for us to make use of this free source of the expensive element, nitrogen; and it makes green manuring with these plants a most important part of farm work, not only as a means of securing nitrogen for the crop itself, but as a source of nitrogen for crops unable to secure it except from soil sources.

An acre of an average crop of red or crimson clover, or of cow peas, will contain one hundred and fifty pounds of nitrogen, equivalent to that contained in fifteen tons of average stable manure. Rye or buckwheat, or other plants, which do not possess this power of securing nitrogen, are much less valuable for this purpose.

Green manuring is particularly useful in the improvement of light lands usually deficient in humus, and in that method of farm practice where exhaustive crops are grown without the addition of yard manure or other forms of organic matter. By the use of the legumes as green manures, and the addition of materials furnishing the essential mineral constituents, potash, phosphoric acid, and lime, light lands may be rapidly improved and made very fertile, while lands used for growing vegetables or fruits may be kept in a high state of cultivation and in good mechanical condition, without the expenditure of money and labor for stable manure, now regarded as so essential by the majority of farmers.

Rye and Buckwheat as Catch Crops. — Rye and buckwheat are of considerable advantage, even though they are able to secure their nitrogen only from soil sources, because their habits of growth permit them to be used as catch crops, or those not interfering with regular rotations.

The addition of the vegetable carbonaceous matter which is contained in these crops is, of course, quite as advantageous as that contained in those having the special power of securing nitrogen; though recent studies of crimson clover show it to be quite as well adapted for a catch crop as those already mentioned, thus limiting the usefulness as green manures of other crops than the clovers or legumes.

Care in the Use of Green Manures. — The turning under of heavy crops of clover or rye in the summer, when the conditions are most favorable for rapid decay, namely, a high temperature, and an abundance of moisture, is sometimes followed by unfavorable results. Whether this is due to the too great development of organic acids from the rapid decay of vegetable matter, as some believe, is not thoroughly established; though it is known that where the soil contains sufficient lime, or when lime is added, the danger in this direction is very much reduced, or altogether obviated.

Green Manures add no Minerals to the Soil.—It is observed, from the foregoing, that the use of green manures can add no minerals to the soil; nevertheless, its chemical qualities are improved. In the first place, constituents existing in the soil, as well as in the subsoil, have been collected by the roots and stored in the whole plant, which, turned under, concentrate these constituents in the surface soil; and, secondly, the constituents contained in combination with vegetable matter are readily given up again, because of the tendency of such substances to decay.

Improvement Due to Lime.—The addition of lime also improves the physical nature of soil. Upon sandy soils its effect is to fill up the openings, which makes them more adherent and more retentive of moisture, thus absorbing less heat during the day, and retaining more at night. On clay soils, the effect of lime is still more important; the fine particles are separated, and the soil made more open, porous, and friable; air and water circulate more freely; the soil is warmer and easier to work.

Tillage.—Natural soils are further improved by tillage. Tillage includes the operations of plowing, cultivating, harrowing, rolling, etc., the result of which is to destroy weeds and foreign growths; to subject larger portions of the soil to contact with air, thus increasing the tendency to decay; and to pulverize the surface soil, and render it more absorptive and porous, and more favorable for the germination of seeds, and for the penetration and activity of the fine roots.

Methods of Plowing.—The methods followed in plowing vary with the conditions and character of the soil. It should be deep enough to include all of the surface soil, and the furrow should be turned in such a manner as to subject the largest surface to the action of the air.

Narrow furrows thrown on edge expose the greatest surface area to the influence of the atmosphere, while a wide furrow, turned nearly flat, presents the least exposed surface. The former method is best adapted for heavy soils, rich in the dormant constituents, and the latter more useful where the object is rather the production of a good tilth or seed-bed.

Proper plowing also greatly assists in surface drainage. The distance between the ridges is called a land, and the narrower the land the better the drainage. Where the natural drainage is good, ridge plowing is not so important. On such lands level plowing is advisable; an even surface possesses many advantages in the cultivating and harvesting of crops.

As a rule, it is not well to bring the subsoil to the surface when the planting of the crop immediately follows the plowing; though on alluvial soils this practice is often followed for the purpose of deepening the surface soils.

Fall Plowing. — Fall plowing is useful in economizing time in the spring, in improving heavy soils, and in destroying many injurious insects. Land plowed in the fall or very early spring is also better able to withstand drouth than if plowed immediately preceding the planting of the crop, particularly if the drouth occurs early in the season.

The gradual deepening of the soil is better accomplished on average soils by deep fall plowing; since a small quantity of subsoil, then brought to the surface, is greatly improved and mellowed by alternate freezing and thawing during the winter.

The cultivation and harrowing of the soil before seeding in the spring should be deep and thorough; all clods should be crushed, and the particles of soil made as fine as possible; the finer the soil is made the more food is made available, and the more moisture is retained. The seed-bed should be deep, clean, and moist, for the proper germination and growth of plants.

Subsoil Plowing. — By subsoil plowing is meant the breaking up of the subsoil, without bringing it to the surface; this is accomplished by a plow of special construction, following in the furrows made by a surface plow. Subsoil plowing is of great importance where the subsoil is hard and compact, and improves the soil, by making the movement of water easier, by admitting the free access of air, and the easy penetration of the roots of plants.

Capillary Attraction. — Water escapes from the surface of soils by means of what is termed "capillary attraction." That is, the interstices, or spaces, between the particles of soil serve as little tubes to conduct the water from the lower levels of soils to the surface, to supply that carried away by evaporation. The coarser the particles of soil and the more porous it is, the larger will be the openings, the less water will be absorbed from the rains, and the more rapid the escape from the surface by evaporation into the atmosphere. In soils the particles of which are not too finely divided and too compact, the reverse is the case. The tubes and pores through which the water passes, if undisturbed, admit of the rapid escape of water; if disturbed, the evaporation is arrested.

Tillage Conserves Moisture. — Cultivating, harrowing, and rolling disturb or break the connection of the pores with the surface, thus reducing the evaporation until the connection of the tubes with the surface is again established. The amount of water transpired by growing plants is enormous — from three to five hundred pounds for each pound of dry matter formed; and its escape, other than through the plant, should be prevented as far as possible.

Tillage also destroys weeds, which require for their growth quite as much plant-food and moisture as cultivated plants.

For cultivated crops frequent tillage is recommended in dry seasons, in order that the greatest possible amount of moisture may be retained where the feeding roots are located; the dry, pulverized surface soil acts as a mulch or blanket, and diverts more of the moisture to the roots of the plant. Too deep cultivation in dry seasons frequently does more harm than good, unless, in the preparation of the seed-bed, the soil has been thoroughly and deeply pulverized.

Chemical Improvement. — Soils are improved chemically by the addition of materials which contain constituents that are liable to be lacking, or which have the power of converting dormant into active constituents. In many cases both of these objects are accomplished at the same time. Materials containing nitrogen, phosphoric acid, and potash are usually regarded as belonging to the first class; while lime itself, and materials containing lime, belong more particularly to the second class. Lime is one of the most useful agents of the farmer, and does not, as is commonly believed, have the tendency to exhaust soils unduly, when its use is properly understood.

Lime acts powerfully upon and hastens the decay of organic matter from both vegetable and animal sources, by virtue of which the nitrogen becomes more quickly available to plants, and, as already stated, lime also assists in the process of nitrification. Lime further aids in liberating potash from insoluble compounds in the soil, thus increasing the store of active plant-food ingredients; it also promotes the formation of compounds with alumina, which have the power of retaining ammonia and potash. The direct effect of lime, as well as the other materials furnishing plant-food, will be discussed in detail in the chapter on manures.

CHAPTER V.

Natural Manures.

A MANURE is, in a broad sense, anything that aids or increases the production of farm crops. Manures may be direct in their effect, by adding to the actual plant-food in the soil, or indirect, by aiding the decay, and making active insoluble plant-food constituents in the soil.

It was shown in previous chapters, that, of all of the constituents which plants need, but four were liable to be exhausted by any system of cropping; these were nitrogen, phosphoric acid, potash, and lime. Direct manures contain one or two or all of these constituents. Anything called a "direct manure," which does not contain one or more of these constituents, cannot add to the stock of true plant-food.

Essential Fertilizing Elements. — Nitrogen, phosphoric acid, potash, and lime are called the essential fertilizing elements, because they are more important in manures than the others that plants require; and a direct manure is useful in proportion to the amount and availability, or direct usefulness, of these constituents contained in it.

Direct manures may also be indirect at the same time; that is, they may contain materials which add no plant-food directly, but which act upon the soil constituents.

Indirect manures are valuable in proportion to the effect which they have upon the soil constituents, and this effect may be due to both physical and chemical causes. Through this distinction, in reference to the action of manures, we are ready to classify them into natural manures and artificial manures.

Natural Manures. — A natural manure is one which may be either direct or indirect, but which has been derived from natural sources, or, in other words, which has not undergone any specific treatment or manufacture. These include all vegetable and animal refuse of the farm and yard, also factory wastes, which contain one or more of the essential constituents. Natural manures are as a rule bulky, and are low grade in the sense that they contain small amounts of the direct plant-food constituents.

Farmyard manure is one of the most important and useful of the natural manures; it is both a direct and an indirect manure: direct, in containing nitrogen, phosphoric acid, potash, and lime, which are actual fertilizing constituents; and indirect, in containing organic or vegetable matter, which aids in the improvement of the physical character of the soil.

It is sometimes called a "general manure," because, as it contains all of the constituents of plant growth, it is liable to be generally useful on all soils.

Farmyard Manure. — Yard manure varies in its composition according to the character of the animals producing it, and the quality of the food, and the object of feeding. Its composition is also influenced by the amount and kind of litter used, and its management after it is secured. The manure from young animals is less valuable than that made when the animals are full grown.

Manure made from fattening animals is richer than that produced by dairy cows; animals fed upon hay and straw furnish manure much less valuable than when the cereal grains constitute a part of the ration.

Manure Produced by Different Animals. — Horse manure is richer in nitrogen, contains less water, and is less variable in composition than that obtained from cows. The manure made from animals consuming rich food is more liable to fermentation than that produced when they are fed upon bulky fodders or watery feeds.

Horse manure is called a "hot manure" because of its tendency to hot fermentation; and is for this reason particularly useful for hot-beds, and for forcing early growth. Cow manure, on the other hand, is called a "cold manure," because less liable to fermentation. Sheep manure contains less water, and is richer in the fertilizing constituents than either horse or cow manure. Pig manure, while quite as watery as cow manure, is richer in nitrogen.

Composition of Stable Manure. — Manure from horse stables in large cities also varies considerably in composition. It contains on the average seventy-five per cent, or fifteen hundred pounds per ton, of water, and twenty-five per cent, or five hundred pounds per ton, of dry matter, which contains all of the manurial ingredients. The water is of no particular value; it simply increases the cost of handling. The dry matter consists of from ten to twelve per cent of ash, and from twelve to fifteen per cent of organic matter. The ash contains from eight to ten pounds each of phosphoric acid and lime, and six to eight pounds of potash; while the organic matter contains from eight to ten pounds of nitrogen.

Its indirect value, however, is often quite as great as, and frequently greater than, its direct value, — first, because of its vegetable matter, which materially improves the absorbing and retaining power of soils; and second, because of the lower forms of life, or bacteria, contained in it, which induce useful fermentations in the soil. Not including the lime, the average ton of city manure contains but twenty-eight pounds of actual fertilizer constituents.

Solid and Liquid Portions. — The nitrogen digested from the food, as well as a large part of the potash, is found in the liquid portions of the manure; while the nitrogen in the undigested portions, as well as a large part of the phosphoric acid, is contained in the solid residue. The nitrogen in the urine is largely in the form of "urea," a compound soluble in water, and is easily decomposed; the potash is also soluble in water. These constituents are, therefore, the most active.

Sources of Loss in Manures. — Manures are susceptible to two direct sources of loss, the first of which is due to fermentation, which results in the loss of nitrogen; and the second is due to leaching, which may finally result in a loss of all the constituents, though it is confined largely to the soluble nitrogen and potash. By fermentation, the nitrogen in the manure is changed to ammonia, usually in the form of a carbonate, which is volatile, and escapes into the atmosphere.

Care of Manures. — Fermentation, causing loss, may be prevented by keeping the manure moist and well packed. The loss through leaching may be stopped if the passage of water through it is prevented. The best method to preserve it is to make it under cover, and in pits made water-tight; by such a method of shelter and protection the maximum amount of manurial value is obtained. The soluble constituents are prevented from being washed into the drain, and the loss of volatile compounds is reduced to a minimum. Where it is not practicable to have water-tight pits, it should be kept in yards that drain to the centre, plenty of absorbent used, and drainage from the roof not allowed to run in the yard; and the product should be removed to the field as often as possible.

Experiments conducted to determine the extent of the loss of valuable constituents due to improper fermentation and to leaching have shown, that, under average conditions of season, the loss from exposure for six months will range from one-third to one-half of the total constituents; this loss falls upon the most active forms, the constituents remaining in the manure after being subjected to such losses are the least active and directly useful.

Manure Preservers. — The loss of ammonia, both in the stables and in manure pits, may also be prevented by the use of land plaster, of kainit, or of superphosphate, which has the power of fixing and retaining the volatile gases. A pound a day per grown animal, sprinkled around in the stable, is sufficient to attain the object. The same proportion and amount may be used on the manure heap. The value of this practice is, however, measured by the care of the manure afterward, since the fixed constituents are still liable to loss from leaching.

The Improvement of Manures.—Manures are improved as they are reduced in bulk, and as the constituents are made available or directly useful; this is accomplished by well-regulated fermentation, or rotting. By well-regulated fermentation is meant that which results in the decay of organic matter with the least loss of nitrogen. The loss from fermentation is greatest when the manure lies in loose heaps, the access of air aiding the decay; the loss is least when it is packed and moist. The mixing of the manures of the various farm animals, hot and cold, also tends to reduce fermentation.

If the fermentation becomes too active, great heat is developed, which causes the rapid escape of moisture; the manure is burned and has a whitish and mouldy appearance,—it is what is called "fire-fanged." Under these circumstances there is frequently a loss of nitrogen. The "fire-fanging" may be prevented by keeping the heap moist.

It is evident, therefore, that the improvement of manures, while it reduces bulk and increases availability of the fertilizing elements, requires care and labor. Whether such improvement will pay or not depends, first, upon the cost of labor, and second, upon the object of use of the manure. Where labor is expensive, and the manure is used for the growing of such gross-feeding field crops as corn, the advantages derived are least. When the handling can be performed by the regular labor of the farm, and where the manure is applied to garden or quick-growing crops, the advantages of such improvement are greatest.

On the whole, however, it is safe to estimate that the least labor necessary to get the manure from the animal to the field is the best policy; that is, while there may be loss, and while the constituents may not be so active, still, the financial results attained are, because of the saving of labor, quite as good.

There is another advantage in the careful fermentation of manures which should not be overlooked, particularly on soils poor in vegetable matter; that is, the development of useful bacteria, the work of which has been already described. What has been said in reference to yard manure is also true for other manures of the farm.

Poultry Manure.—This is richer in all of the essential elements than any other natural manure of the farm. It contains less water, and is not so liable to hot fermentation if kept moist.

Application of Yard Manure.—Two points should be kept in mind in the application of yard manures,—first, that they are essentially nitrogenous products; and second, that they are particularly valuable because of the useful ferments contained in them. If too much is added at one time, a loss of nitrogen is liable to follow, and the benefits derived from the ferments are limited to small areas. The manure of the farm should be distributed as far as possible, and supplemented by more concentrated materials. Coarse manures are better adapted for heavy lands, while those which are well rotted are more useful on light soils.

Composts. — In addition to the yard manure, there are about most farms wastes of considerable importance, weeds, grasses, and coarse growths of any kind, which all contain greater or less amounts of the manurial constituents. These should be carefully utilized, and may be profitably used as absorbents in the barnyard. When this method is adopted, the weeds should be cut before they have matured, or they furnish an excellent means of transmitting foul seeds. These waste products may also be used in making what are called "composts." These, of course, differ according to the conditions of the farmer. Where peat or muck is available, they are more advantageous than where such products are not at hand. The main object of the compost heap is to cause a more rapid decay of such products, and without the loss of essential constituents.

A good compost heap may be made by placing a layer of manure, then a layer of the weeds or waste products of any kind, then a layer of lime or ashes, the whole well moistened, and the order repeated until all of the products are used. The manure starts fermentation, the lime aids in the rotting, as well as to prevent acidity and to keep the heap alkaline, and the moisture prevents too hot fermentation. By careful management destructive fermentation is prevented, the bulk is very materially reduced, and the quality of the constituents greatly improved. The chief difficulty in the making of composts, as well as with other methods used in the improvement of manures, is the expense of labor.

It pays to take good care of, and to save, manurial products, and to see to it that wastes are reduced, and the improvement of the quality of the constituents by the methods suggested is frequently of considerable financial advantage.

Muck, or Peat. — On many farms there are low, wet places where the conditions are favorable for the collection of partially decayed vegetable matter. The material thus formed is called "muck," or "peat." The thickness of the deposit and its character depend upon the time during which it has been formed, and the character of the climate.

Muck is used mainly as a source of humus, and as an absorbent for use in stables or yards. Fresh muck contains on the average seventy-five per cent of water, three-tenths per cent of nitrogen, and traces of potash, phosphoric acid, and lime. Air-dry muck contains on the average twenty-one per cent of water, one and one-third per cent of nitrogen, one-tenth per cent each of phosphoric acid and potash, and nine-tenths per cent of lime. The value of the muck as a source of humus is measured by its content of nitrogen, while its value as an absorbent depends upon its content of organic matter. The usefulness of muck for either of these purposes is further modified by the labor necessary to secure it in a dried condition.

The usual method of procuring it is to throw it out of the bed into heaps, and allow it to dry before it is used either upon the fields or in the stables. Where a muck bed exists upon a farm, it should first be studied in reference to its possible drainage. If it can be drained, it is liable to prove more useful where it lies than for the other purposes mentioned; since soils rich in peaty matter are particularly valuable, when properly managed, for growing onions, celery, and potatoes. Large areas of peaty soils in this country, that have been properly drained, are now devoted to these crops; before draining they were absolutely valueless.

In addition to the natural farm wastes, farmers frequently have easy access to certain factory wastes. These may be divided into two classes, nitrogenous and potassic. Of the nitrogenous materials, wool and hair wastes are probably the most important. These are very rich in nitrogen; both are, however, usually mixed with other materials, and vary widely in their composition.

Wool Waste. — Wool waste contains on an average ten per cent of water, five and one-half per cent of nitrogen, one per cent of phosphoric acid, and two per cent of potash.

Hair Waste. — An average analysis of hair waste, as determined at the New Jersey Experiment Station, shows it to contain thirty-two per cent of water, seven and two-tenths per cent of nitrogen, and eight-tenths per cent of phosphoric acid.

Felt Waste. — Felt waste is similar to wool waste in that its nitrogen is contained in the wool, though variable in composition on account of the varying proportions of cotton used in its manufacture. Analyses show it to contain about eight per cent of nitrogen.

Leather Meal. — Leather meal is a product found in considerable quantities in towns where the manufacture of shoes is an industry. It contains on the average ten per cent of water, and seven of nitrogen. The nitrogen in the leather meal is even less available than in the other products mentioned, because the leather has passed through a process, the very purpose of which was to make it less liable to decay.

The purchase of these materials is only advisable when they can be procured very cheaply. Their application is useful when the object is gradual increase in fertility, rather than immediate increase in crop. Hair wastes have been found advantageous in the growing of berries, hops, and other slow-growing crops, while wool and leather have materially improved meadows and permanent pastures. The nitrogen in these materials is improved in form, and made more quickly available, when composted with manure.

Wood-Ashes. — Of the potassic manures, unleached wood-ashes are the most useful. The pure ashes from the different varieties of wood vary in composition; as a rule, the softer woods contain less, and the hard woods more, potash, the range being from sixteen to forty per cent.

Ashes also contain lime in large amounts, while phosphoric acid is contained in much smaller quantities. Wood-ashes, as usually gathered for market, however, contain very considerable portions of moisture, dirt, etc., which cause a variability in composition not due to the character of the woods from which they are derived. The average analysis of commercial wood-ashes shows them to contain less than six per cent of potash, two per cent of phosphoric acid, and thirty-two per cent of lime. Leached wood-ashes contain on the average thirty per cent of moisture, one and one-tenth per cent of potash, one and one-half per cent of phosphoric acid, and twenty-nine per cent of lime.

Ashes are probably one of the best sources of potash that we have, so far as its form and combination are concerned, being in a very fine state of division, and in such a form as to be immediately available to plants. Ashes also have a very favorable physical effect upon soils, the lime present, of course, aiding in this respect. Canada is now the main source of wood-ashes, the substitution of coal for wood making the supply in this country for commercial purposes very limited. Owing to the variability of this product, it should always be bought subject to analysis, and to a definite price per pound for the actual constituents contained in it; which should not be greater than the price at which the same constituents could be purchased in other quickly available forms.

Marl. — Marl may contain one or more of the constituents, phosphoric acid, potash, and lime. Shell marls are usually very rich in lime, but contain only traces of phosphoric acid and potash. The green sand marls of New Jersey often contain very considerable amounts of phosphoric acid and potash, though they vary widely in composition. They contain on the average two and two-tenths per cent of phosphoric acid, four and seven-tenths per cent of potash, and two and nine-tenths per cent of lime. These constituents, particularly the potash, are, as a rule, slowly available.

Marl, however, is an important amendment to soils, not only because of its content of mineral constituents, but because these constituents are associated with products that have a very favorable mechanical effect upon soils. Large areas of land in the State of New Jersey formerly unproductive, chiefly because of physical imperfections, have been made very productive mainly through the application of marl.

The use of marl is now less general than when the fertilizing constituents from artificial sources were dearer, and when the labor of the farm was more abundant and cheaper. The quicker effect of more soluble fertilizer constituents has had an influence in reducing the use of marl where quick returns are desirable. Where farmers have deposits upon their own farms, or within short distances of them, and can secure it at a low price per ton, it is a desirable method of improving land.

The results from the use of marl are frequently due quite as much to the improvement given to the physical condition of soils as to the increase in fertility furnished by the essential mineral constituents. Marls may be carted and spread upon the land when other work of the farm is not pressing, thus making it possible to get a considerable addition of fertility at a small expense.

Lime. — Lime, while an essential constituent of plants, is usually more abundant in soils than the other mineral constituents, phosphoric acid and potash. It is, however, regarded as a direct source of plant-food in a great many cases, though its greatest value lies in its favorable action upon soils. This action is both physical and chemical, and has already been discussed in a previous chapter.

Lime, as is it generally understood, is an oxide of calcium, and is produced by burning limestone, or carbonate of lime. The lime loses the carbonic acid when burned in the kilns, and the oxide of lime remains behind; this is usually termed "quicklime." The quicklime, before it is applied to the soil, is usually slaked; this is done by adding water, which the lime absorbs, and falls to a powder. Slaked lime, also called caustic lime, is a calcium hydrate.

The more completely limestone is burned, the better the quicklime, and the more completely it slakes. We have, when we speak of lime, three forms: limestone, quicklime, or burned lime, and slaked lime, each differing from the other in composition.

Quicklime absorbs moisture, and slakes when exposed to the atmosphere. Lime thus slaked is called "air-slaked lime," and is usually less completely changed to a hydrate than when water is added. Quicklime also absorbs carbonic acid from the air, and changes back to the limestone form. Lime in the carbonated form, if finely pulverized, is better for liming light lands than the caustic lime; for heavy lands, caustic is preferable to the carbonate.

What is termed "marble lime" is made from pure limestone. What are called "limestones" frequently contain considerable magnesia, in which case they are termed "magnesian limestones." The larger number of the limestones of New Jersey are of this class; they contain from fifty to sixty per cent of calcium oxide, and thirty per cent or over of oxide of magnesia.

Oyster shells are nearly pure carbonate of lime; oyster-shell lime, though containing no magnesia, is usually mixed with more or less dirt and other impurities, and is therefore not as rich in lime as that derived from pure limestone.

Gas-Lime. — The lime from gas-works is also frequently used as manure; in these works quicklime is used for removing the impurities from the gas. Gas-lime, therefore, varies considerably in composition, and consists really of a mixture of slaked lime, or calcium hydrate, and carbonate of lime, together with sulphides and sulphites of lime. These last are injurious to plant life, and gas-lime should be applied long before the crop is planted, or at least exposed to the air some time before its application; the action of air converts the poisonous substances in it into non-injurious products. Gas-lime contains on an average forty per cent of calcium oxide.

Gypsum or Land Plaster. — Gypsum is a sulphate of lime containing water in combination. Pure gypsum contains thirty-two and one-half per cent of lime, forty-six and one-half per cent of sulphuric acid, and twenty-one per cent of water.

Plaster of Paris is prepared from pure gypsum by burning, which drives off the water it contains. Gypsum, like other forms of lime, furnishes directly the element calcium, and also exerts a favorable solvent effect upon the soil. It was formerly used in large quantities, particularly for clover; and it is believed that its favorable effect was due, not so much to the direct addition of lime, as to its action upon insoluble potash compounds in the soil, in setting free potash. Thus the application of plaster caused an increase in crop because of the potash made available.

We have in the Eastern States two main sources of gypsum, namely, Nova Scotia and Cayuga, N.Y. Nova

Scotia plaster is purer than that obtained from New York. The New York plaster, however, frequently contains appreciable amounts of phosphoric acid.

Salt. — Common salt is sometimes used as a manure. It supplies no essential plant-food constituents; and its value is still a disputed point, though it is admitted that, where its use is favorable, it is due to indirect action in aiding the decomposition of animal and vegetable matter, increasing the absorbing power of soils, and, by its reaction with lime, acting as a solvent for phosphates. Salt is frequently applied in connection with nitrate of soda for wheat crops, to prevent a too rapid growth of straw.

The Application of Lime. — The quantity of lime to be applied may vary according to circumstances; heavy lands, rich in organic matter, may receive more, and lighter lands, less. The usual amount in the Eastern States, on average land, ranges from one to three tons of quicklime per acre. This is applied once in six or seven years, the application of small quantities being frequently more useful than large quantites applied at wider intervals. Lime should be applied on the surface, as its tendency is to work into the soil, and gradually get below the surface soil. The time of application, also, varies with the kind of crop and the character of the soil. For pasture-lands or mowing-fields, the early spring or fall are the best seasons to apply it. It should never be used directly with commercial fertilizers containing ammonia and soluble phosphoric acid, as it sets free the ammonia, and reduces the solubility of the phosphates.

CHAPTER VI.

Artificial and Concentrated Manures; Nitrogenous Materials.

As farm lands become exhausted of their essential plant-food constituents by the continual sale of crops, the manures available to the farmer, both from the natural wastes of the farm and from such materials as lime, ashes, etc., are often insufficient to keep up their original fertility. At the present time, too, the tendency of farming in this country, especially in the Eastern States, and in the vicinity of large cities, is toward special crop farming, which requires that soils should be abundantly supplied with active plant-food.

These conditions have caused a rapid development of the sources of supply of suitable materials that furnish the constituents liable to be lacking, or contained in too small amounts, in the soil; viz., nitrogen, phosphoric acid, and potash.

Classes of Materials. — These materials are divided into three distinct classes, — namely, nitrogenous, furnishing nitrogen; phosphatic, furnishing phosphoric acid; and potassic, furnishing potash.

Manures made from these materials are called "artificial," "concentrated," or "commercial." They differ from the natural manures mainly in being more concentrated, though frequently the constituents in them are more immediately available to the plant. The manurial elements, if in a form in which plants can use them, are quite as much actual plant-food when contained in these materials as when furnished by the more familiar natural manures.

Natural manurial products, or homemade materials, are used in their original state or applied directly to the soil. Artificial products, as a rule, require treatment previous to their use.

Nitrogen. — Nitrogen is the most costly element of manures. It is absolutely essential to all organized life, whether animal or vegetable; it is the basis of the albuminoids of plants, the casein of milk, and the fibrin of blood. Nitrogen occurs in three forms, and all these forms exist as commercial manure products. The form means its combination with other chemical elements; namely, nitrogen as nitrates, nitrogen as ammonia, and nitrogen as organic matter.

Forms of Nitrogen. — Nitrogen in the form of a nitrate means its combination with oxygen in such proportions as to form nitric acid, united with a base like soda or potash; thus, we have nitrates of soda, potash, lime, etc.

Nitrogen as ammonia means its combination with hydrogen in such proportion as to form ammonia. Ammonia gas consists of one part of nitrogen and three of hydrogen. This gas readily combines with various acids, as sulphuric, nitric, etc., to form ammonia salts.

Nitrogen in the form of organic matter means its combination with the chemical constituents, carbon, hydrogen, and oxygen, either as animal or vegetable substances.

A large number of materials, differing widely in their character and composition, contain nitrogen in this form.

Plants that derive their nitrogen from the soil absorb it chiefly in the form of a nitrate; when nitrogen is applied in this form, no changes are required to enable it to serve as a direct food. Materials furnishing nitrates are, therefore, regarded as of the greatest importance in the manufacture of commercial manures.

Ammonia, while it does nourish plants directly, usually undergoes a change into nitrate first, though this change proceeds rapidly when the conditions mentioned as favorable for nitrification are present. As a rule, therefore, an appreciable time does elapse before all the nitrogen in ammonia serves as plant-food.

Nitrogen in organic forms is first changed by the decay or rotting of the substance into ammonia, and the ammonia is then changed into a nitrate. The rapidity of this decay depends both upon the character of the substance itself, and upon its physical form and its mechanical conditions or fineness of division. The tougher and more dense the substance, and the coarser the particles, the longer the time required to rot, and the more slowly available as nitrogenous food. The rapidity with which nitrogen may become useful as food to plants is, therefore, determined by its chemical form.

Nitrates, since they are immediately useful to the plant, may all be absorbed by the crop upon which they are applied, while ammonia salts and organic nitrogen may be only partially used, because the necessary changes for them to undergo may not take place completely before the plant is fully matured.

The different results obtained from the use of the different forms of nitrogen determine what is called its "agricultural value," or the improvement it causes in the growth of the plant.

This agricultural value, which is true of any manure, is, too, separate and distinct from the commercial value, or cost in market; which is determined by market and trade conditions, as cost of production, transportation, selling, and the demand for it in other industries. It is for this reason that the best forms of plant-food may, and frequently do, cost less per pound of the actual ingredient than when furnished by other more slowly available, and less directly useful, forms.

Nitrate of Soda. — Nitrate of soda, also called Chile saltpetre, is the chief source from which nitrogen as a nitrate is secured for manurial purposes. It possesses chemical and physical properties which distinguish it from all other materials; it is a salt with a definite chemical composition. When pure it contains sixteen and forty-seven hundredths per cent of nitrogen.

Vast natural deposits of the crude nitrate salts occur in the rainless districts of South America, though mainly in Chile. The crude salts are relatively poor, and also variable in their content of nitrogen; hence before shipment they are purified by dissolving in water and recrystallizing, the impurities remaining being chiefly water and ordinary salt.

The commercial product is quite pure, containing on the average sixteen per cent of nitrogen. It resembles in appearance ordinary salt, though the use of salt as an adulterant has not been practised to any extent in this country. It is completely soluble in water, diffuses readily throughout the soil, and, since it forms no insoluble compound with soil constituents, is liable to be washed into the lower layers of the soil, and finally into the drains, if applied in the fall of the year, or in too large quantities. It is very beneficial for early and quick-growing crops, particularly upon light sandy soils, because it is ready for use as soon as applied.

The soda with which the nitrate is combined does not possess any value as a plant-food constituent, though it is believed to exert a beneficial effect upon the physical character of soils. This value is not taken into account in commercial transactions.

The use of nitrate is rapidly increasing where its advantages are well known.

Nitrate of potash, or saltpetre, is another nitrate salt used to some extent, though its cost of production is too great to admit of its competition with the nitrate of soda. It is a concentrated product, and furnishes potash in addition to the nitrogen.

Sulphate of Ammonia. — Ammonia for fertilizing purposes is secured almost entirely from sulphate of ammonia, which is another chemical salt of definite composition, and is one of the most concentrated forms in which nitrogen occurs. It contains, when pure, twenty-one and two-tenths per cent of nitrogen. It is derived chiefly as a by-product from the manufacture of gas by the dry distillation of coal, and is also secured in smaller quantities in the manufacture of bone-black for sugar refineries, and in the distillation of refuse animal matter.

The nitrogen contained in these products is, by the

process of distillation, viz., in the absence of air, driven off in the form of ammonia gas, which is conducted into receptacles containing sulphuric acid, which fixes the ammonia in the form of a sulphate. This is a crude product, and is usually dissolved, recrystallized, and purified, and is then reasonably uniform in composition, and contains on the average twenty and one-half per cent of nitrogen. When sold without this method of purification, it is called "brown sulphate of ammonia;" this is more variable in composition, with moisture, and insoluble and oily matters as impurities. Brown sulphate often contains less than eighteen per cent of nitrogen. Sulphate of ammonia, while freely soluble in water, is readily absorbed by the soil, and may be applied in the fall without danger of serious loss, and is, next to nitrate of soda, one of the best forms of nitrogen for fertilizing purposes. It is a particularly useful form on clay and clay loam soils.

Nitrates and ammonia salts possess two chief advantages; viz., solubility, hence ease of distribution in the soil, and availability, or quick usefulness to the plant.

The cost per pound of nitrogen in the form of ammonia is usually greater than for the nitrate of soda, because of the comparatively limited supply, and because more largely used in the arts.

Organic Nitrogen. — Organic nitrogen is obtained from a wide variety of sources, and is composed of both animal and vegetable matter. The commercial materials, from whatever source derived, unlike the salts mentioned, are not definite chemical compounds; those of the same name, even, vary in their composition, and also in their agricultural value or usefulness to plants, largely according to their method of preparation for market.

Dried blood, for instance, is rich in nitrogen when carefully prepared, though commercial samples show wide variations in composition, and in the quality of the nitrogen. A pound of nitrogen, therefore, may have a very different value in one sample than in another, under uniform conditions of use, while under the same conditions of use a pound of nitrogen in the form of nitrate is equally valuable from whatever source derived.

Materials containing organic nitrogen are not soluble in water, and the nitrogen is not immediately available to plants. Their value as a source of this element is, therefore, measured by conditions which favor the rapid change of their nitrogen into soluble and available forms. The first condition is fineness of division, which permits of a more even distribution, and the exposing of a larger surface area to the action of the agencies in the soil and air; and the second is the physical character of the material itself. If it is hard and dense, decay will be slower, since the processes which cause it are resisted; if soft and porous, the rotting is more rapid, because the agencies which cause it are encouraged to act.

Dried Blood. — Dried blood is one of the most valuable sources of organic nitrogen. Its fineness of division permits of its easy and uniform distribution, and its physical character is such as to encourage rapid decay under suitable conditions.

What is called "high-grade blood" is red in color, and is quite uniform in composition, ranging from twelve to fourteen per cent of nitrogen. This form of blood is used largely in the arts, hence the supply for manurial purposes is limited: lower grades are in greater supply; these are darker in color, vary widely in composition and physical character, and frequently contain, as impurities, moisture, hair, bone, etc.; the two latter, to be sure, contain nitrogen, but in forms less useful than in the blood.

Dried Meat, or Azotine. — This material is probably, next to blood, the most important source of organic nitrogen. It is obtained by separating the meat from the bones, extracting the fat by steaming or solvents, and drying and grinding into powder. It is usually in excellent mechanical condition, and when free from bone is quite uniform in composition, containing on the average twelve per cent of nitrogen.

Tankage. — Tankage is the dried refuse from slaughter-houses and butcher-shops, and is composed of various wastes — as offal, skin, bone, hair, and meat; it contains both nitrogen and phosphoric acid, though usually classed as a nitrogenous product.

The source from which this material is derived indicates at once that it must be variable in composition; it shows wide ranges in its content of both nitrogen and phosphoric acid. These variations in composition are largely due to the quantity of bone contained in it. The larger the proportion of bone, the lower the percentage of nitrogen, and the smaller the proportion of bone, the higher the content of nitrogen. What is called "concentrated tankage" is made up more largely of extractive animal matter, and is more uniform in composition, and much richer in nitrogen.

The lack of uniformity in both the chemical composition and physical character of tankage, makes it impossible to give to it a distinct position as a fertilizing product, since the results derived from its use under uniform conditions in other respects must naturally be variable.

Dried Fish. — Dried fish was formerly an important and considerable source of organic nitrogen. It is derived mainly from the waste resulting from the extraction of the oil from the menhaden, a fish not used for food, though valuable for this purpose. The waste from fish-canning establishments also furnishes a considerable amount of this product.

The "menhaden pomace" is rich in quickly available nitrogen, and also contains considerable phosphoric acid, derived from the bone contained in it; it is reasonably uniform in composition. The waste from the canneries contains more of the skin and bone, and is more variable in composition. Aside from the source from which it is derived, the chief cause of variability in composition is the content of water. Frequently, too, acid is used to prevent decomposition, which, while useful in making the constituents more available, renders the product less concentrated and less easily dried. Well-dried samples contain on the average seven to eight per cent of nitrogen, and the same of phosphoric acid. Finely ground, it decays rapidly in the soil, and is highly regarded as a manure.

Fish was one of the first materials used as a fertilizer. The custom in this country in early times, of using a fish in each hill of corn, is still practised in regions near the sea where they are easily procured.

Leather Meal. — This material, already described in a previous chapter, is frequently treated by various processes, in order to render its nitrogen more available. Chief among these are heating to a high temperature, and steaming, which change its original form and structure, making it mealy and crumbly. Methods of treating with borax and benzine, and dissolving in sulphuric acid, are also practised. Experiments have shown that all of these processes aid materially in improving the quality of the nitrogen in this product.

Horn and Hoof Meal. — These materials are rich in nitrogen, and are quite uniform in composition, though in their original condition they are slow to decay. They are frequently treated in somewhat the same manner as leather, in order to render the constituent nitrogen more directly available to plants.

Thus, while these products, containing a high content of nitrogen in very slowly available forms, do have some value, it is doubtful whether their use, either in their original or treated form, is advisable, except when they can be purchased at a much lower price per pound than in forms of known value.

Cottonseed Meal. — A few vegetable waste products contain sufficient nitrogen to warrant their use as concentrated manures. Among these cottonseed meal is probably used to the greatest extent. The cottonseed is first hulled, ground and steamed, and the oil extracted. It is quite uniform in its composition, and contains on the average six and eight-tenths per cent of nitrogen. Its value as a food for stock, however, limits its use as a fertilizer at points distant from place of production. It is fine and dry, and decays rapidly in the soil, and is regarded as one of the best forms of organic nitrogen.

Castor Pomace. — Castor pomace is similar to cottonseed meal, both in its composition and in the quality of its nitrogen. It is the refuse castor bean after the oil has been extracted. It is not an important source of organic nitrogen, though practically the whole product is used as a manure.

Organic Nitrogenous Materials are Variable. — As has been indicated, organic nitrogenous materials are, on the whole, variable products, both in respect to their content of nitrogen, and to its availability. Those that are uniform in respect to their composition are more valuable than the others, since their availability may be tested, and an average arrived at.

Those high-grade products, like dried blood, azotine, dried fish, and cottonseed meal, which are fairly uniform in these respects, show, both by chemical and field tests, a high percentage of availability, which does not vary greatly with different samples; while those like tankage sometimes show a high and sometimes a low availability, because of the lack of uniformity in the proportions of their component parts.

The Use of Nitrogen. — Great care should be exercised in the purchase and use of nitrogen, first, because it is an expensive element; and second, because when it is in a form useful to plants it is entirely soluble in water and freely movable, and, therefore, liable to be washed away and lost. The other elements, phosphoric acid and potash, cost much less than nitrogen, and are fixed in the soil, and, as a rule, are taken out of the soil only by the plants themselves. It is estimated that, even when the greatest care is exercised, not more than two-thirds of the nitrogen applied as manure is used by the crop. Carelessness in its use results, of course, in much greater losses.

Application of Nitrates. — Nitrates being completely soluble, should not be applied in large quantities in the fall of the year, or in the early spring before vegetation begins. The most economical use of this form of nitrogen lies in its fractional application to growing crops in quantities sufficient for their needs. An overabundance of available nitrogen frequently causes a too rapid development of leaf. It should be applied when the foliage is dry, either preceding or following a rain, in order to effect its solution, unless it is cultivated into the surface soil. The favorable effect of nitrates applied in this way is very quickly noticeable, especially upon vegetable and garden crops.

The above is true, though in a less degree, of ammonia salts. Frequently losses occur through too heavy applications at the wrong time.

Application of Organic Nitrogen. — Organic forms of nitrogen may be applied at any time, and in larger quantities. The more insoluble materials should be applied in amounts known to be in excess of the needs of the crop; since, even under the best conditions, the nitrogen contained in them is slowly available.

Materials like blood and fine-ground fish will rot completely in an average season; while horn, hoof, hair, leather, wool, etc., may require several seasons to effect their complete decay.

CHAPTER VII.

Artificial and Concentrated Manures; Phosphates.

The phosphoric acid in artificial manures is derived from compounds called "phosphates." In phosphates the phosphoric acid is united with lime, iron, and alumina, forming phosphates of lime, iron, and alumina, as the case may be. The phosphates of lime are better calculated for the purpose, and are, therefore, used more largely than any other as a source of phosphoric acid in the manufacture of artificial manures.

The phosphates available for this purpose are not, however, pure salts, but exist in combination either with organic substances, or with minerals, or both; the content of phosphoric acid and its combination with other substances determining the usefulness of the phosphate to the manure-maker.

The phosphoric acid in these materials is difficultly soluble in the soil water; and hence in their original condition, or in the crude raw forms, they give up this element in proportion as they decompose or decay in the soil. Those in combination with organic substances, either animal or vegetable, are, as a rule, more quickly useful as a source of phosphoric acid than those composed entirely of mineral constituents.

Animal Bone. — The bones of animals are the chief source of phosphates that exist in combination with organic matter, and were for a long time the main source for manurial purposes.

Bone consists chiefly of three classes of substances; viz., moisture, organic matter, containing nitrogenous and fatty matter, and phosphate of lime, — the proportion, particularly of the nitrogen and phosphoric acid, depending upon the kind of bone, and the method of its treatment.

Bone from the same kind of animal differs in composition according to the age of the animal, and according to its location in the body. In a general way the younger the animal the softer the bone, the poorer in phosphate of lime, and the richer in nitrogen; the older the animal, the richer in phosphate of lime, and the poorer in nitrogen. The large and hard thigh bones of an ox, for instance, differ in composition from the softer and more porous bones of other parts of the body.

Treat a bone with dilute hydrochloric acid, and you dissolve the phosphate of lime, and leave the soft pulpy animal matter, which retains its original shape. Burn the bone, and you drive off the organic matter, and leave the porous phosphate of lime in the original shape, showing the structure of the bone. The phosphate of lime of the harder bones is dense and compact; that from the softer bone is more open and porous. The chief cause of variation in the composition of bones used as manure, however, is due to the treatment they receive. This is recognized by manufacturers and dealers, and different names of brands are used to indicate the method of manufacture or treatment; as applied, however, they do not always correspond to the methods of treatment.

Raw Bone. — The term "raw bone" is properly applied to bone that has not suffered any loss of its original constituents in the processes of its manufacture; and is for this reason highly regarded by farmers, who believe that it is purer than any other form. This is true in a large measure, though the fact that it is raw bone is not altogether an advantage from the standpoint of usefulness. Raw bone too often contains considerable fatty matter, which makes it a difficult process to grind it fine, and which also has a tendency to retard the decay of the bone in the soil. A considerable amount of fat also reduces proportionately the percentage of the valuable constituents, phosphoric acid and nitrogen. Good raw bone, free from meat and excess of fat, should contain on the average twenty-two per cent of phosphoric acid, and four per cent of nitrogen.

Fine Bone. — The trade terms "bone meal," "bone dust," and "fine bone" are used to indicate mechanical condition, or fineness of division, and do not refer especially to composition. These names should not be taken as indicating the fineness without personal examination, since frequently the products do not, in this respect, correspond to the name. Fineness is an important consideration, since, the finer the bone, the quicker it will decay, and its constituents become available to plants.

Boiled and Steamed Bone. — The larger portion of the bone used as manure has been boiled or steamed for the purpose of freeing it from fat and nitrogenous matter, both of which are products valuable for other purposes. The fat is, of course, of no value as a manure, and its absence is an advantage. The nitrogen, while useful as a manure, is extracted chiefly for the purpose of making glue and gelatine.

By boiling or steaming, the bone suffers a loss of its original constituents, the chief result of which is to change the proportions of the nitrogen and phosphoric acid contained in it. Steamed or boiled bone contains more phosphoric acid, and less nitrogen, than raw bone, and is also more variable in composition, the relative percentage of these constituents depending upon the degree of steaming or boiling to which the bone has been subjected.

Bone that has been used for the purpose of making glue, where the chief object is to extract the nitrogenous matter, contains from twenty-eight to thirty per cent of phosphoric acid, and from one and one-quarter to one and three-quarters per cent of nitrogen. The steaming of bone, particularly when conducted at high pressure, also exerts a favorable effect upon the physical and mechanical character of the bone. It destroys its original structure, makes it soft and crumbly, and often reduces it to a finer state of division than can be readily accomplished by grinding; and, since it is also free from fat, and is finer, it is more directly useful as a source of phosphoric acid to plants than purer raw bone.

Experiments have shown that the phosphoric acid in fine steamed bone may all become available in the soil, under average conditions, in one or two seasons; while that in the coarser, fatty raw bone is not completely used in three or four years, and sometimes longer.

In some cases, the fat is extracted from bone by means of such solvents as petroleum or benzine. These methods of extracting the fat have the advantage of increasing the relative proportion of the nitrogen, this element not being attacked by the solvents.

The more complete extraction of the fat and moisture by these methods also aids in the final preparation of the bone by grinding. Bone prepared in this way frequently contains as high as six per cent of nitrogen, and twenty per cent of phosphoric acid.

The nature and composition of animal bone is such as to make it a valuable source of phosphoric acid; and, while it is largely used with nitrogenous and potassic materials in the manufacture of artificial manures, its best use is, perhaps, in the fine ground form, particularly for soil improvement and for slow-growing crops.

Phosphoric acid applied in this form gradually gives up nitrogen and phosphoric acid to the plant; and its physical and chemical conditions are such that it forms in the soil, during the growing season, no compounds more insoluble than the bone itself. Of all the phosphatic materials available as manure, bone is the only one that is now used to any extent without further treatment than simple grinding.

Bone-black or Animal Charcoal. — This material becomes an important source of phosphoric acid for artificial manures after it has served its chief and first purpose in clarifying sugar. In making bone-black only the best bones are used; they are cleaned and dried, and placed in air-tight vessels, and heated until all volatile matter is driven off; the resultant product, which retains in part the original form of the bone, is then ground to a coarse powder; it then becomes a bone charcoal, consisting chiefly of carbon and phosphate of lime, though also containing small amounts of magnesia and carbonate of lime.

Bone-black, as received from the refineries, contains the impurities gathered there, consisting chiefly of vegetable matter and moisture. It is somewhat variable in composition, containing from thirty-two to thirty-six per cent of phosphoric acid and a small amount of nitrogen. It decays slowly in the soil, and is not now used to any extent directly as a manure.

Bone-ash. — Bone-ash, though not a large, is an excellent source of phosphoric acid. It is exported in considerable quantities from South America, where the bones are burned, and the bulk reduced, in order to facilitate transportation. It does not contain nitrogen, and is more variable in composition than bone-black, though usually somewhat richer in phosphate of lime. Good samples contain from twenty-seven per cent to thirty-six per cent of phosphoric acid.

Bones themselves, and the phosphates derived from bones, constitute a class differing from other phosphates used in making manures, in that they are derived directly from organic materials; and, as a class, they possess characteristics due to this fact, which render them more useful than those derived from purely mineral sources.

Mineral Phosphates. — These constitute a class of products differing from those of immediate or recent animal origin, mainly in the fact that they are not combined with organic matter, and are more dense and compact in their structure. They occur in several different forms, and are procured from distinct sources.

South Carolina Rock Phosphates.—These are found both on the land and in the beds of rivers in the vicinity of Charleston, S. C., and are sometimes called "Charleston Phosphates." The deposits vary in thickness from one to twenty feet, through which the phosphate is distributed in the form of lumps or nodules, ranging in weight from an ounce to over a ton. These nodules are irregular, non-crystalline masses, often full of holes, which contain clay or other non-phosphatic materials. That obtained from the river is called "river phosphate," or "river rock;" and that from the land, "land phosphate," or "land rock." The two varieties do not differ materially in composition, particularly in the content of phosphoric acid.

The rock contains from twenty-six per cent to twenty-eight per cent of phosphoric acid. Its uniformity, in connection with the fact that it contains but small percentages of compounds of iron and alumina, minerals which prevent its best use by the manufacturer, makes it a highly satisfactory source of phosphoric acid.

The river rock is secured by dredging; that from the land is largely dug. In either case, it is washed to remove the adhering matter, and then dried, when it is ready for grinding or shipment. South Carolina rock phosphate, when very finely ground, is called "floats." It is sometimes used upon the land in this form.

These deposits were first worked in 1868, though the presence of phosphate at this point was known at a much earlier date.

Florida Phosphates.— The presence of phosphate in commercial quantities in Florida was discovered in 1888, since which time very great progress has been made in developing the deposits. These deposits occur in a number of forms, — first, "soft phosphate," a whitish product, somewhat resembling clay, and largely contaminated with it; second, "pebble phosphate," consisting of hard pebbles, occurring both in river-beds and upon the land, and mixed with other materials; and third, "rock," or "bowlder phosphate," which occurs in the form of stony masses, or bowlders, both large and small. These three forms also differ widely in composition, both in reference to their content of phosphoric acid and in respect to the presence of other minerals.

The soft phosphate is the poorest in phosphoric acid: it is easily prepared, and is largely used directly upon the land; it is also the most variable in composition, ranging from eighteen to thirty per cent. The pebble rock is also variable in composition, though, when washed free of sand and clay, it is richer in phosphoric acid than the soft variety; good samples contain as high as forty per cent and over of phosphoric acid. The bulk of the "Florida Phosphate" is believed to exist in the pebble form.

The rock or bowlder phosphate, though apparently much less in amount, is more uniform in composition, and is much richer than either of the other forms. The clean, dry bowlder phosphate often contains as high as forty per cent phosphoric acid, far exceeding in richness the South Carolina rock superphosphate.

Canadian Apatite. — This material is a crystallized rock of true mineral origin, and occurs associated to a greater or less extent with other materials. It is, therefore, not uniform in character, the phosphoric acid varying according to the amount of the other substances present.

It is mined in the provinces of Quebec and Ontario, and separated into various grades at the mines. The mining is expensive, and the necessity for grading in addition makes the cost of production proportionately high. The highest grade of this phosphate is very pure, containing forty per cent of phosphoric acid.

Iron Phosphate. — This is a waste product from the manufacture of steel from phosphatic iron ores, by what is known as the "basic process." It is sold under several names, as "Thomas Phosphate Meal," "Phosphate Slag," "Basic Slag," and "Odorless Phosphate." It is produced in large quantities in England, France, and Germany; and in those countries is not only one of the cheapest sources of phosphoric acid, but is regarded as a very valuable product. It is not produced to any extent in America, is known under the name of "Odorless Phosphate," and is not largely used. It contains from fifteen to twenty per cent of phosphoric acid, in the form of phosphate of lime, in connection with large amounts of lime and oxide of iron. It is used almost altogether in the form of a fine powder, since it is not suitable for the purposes of the manufacturer. When very finely ground, the phosphoric acid is quite as active as that contained in fine bone meal, and is especially suitable for clay and sandy soils and for meadows.

Phosphatic Guanos. — Previous to the discovery of the phosphates in South Carolina, these guanos were a very important source of phosphoric acid; they are now but little used in this country. They are obtained from the rainless districts of the world, chiefly from the islands bordering the coast of South America and from the West Indies. They are derived from the excrements of birds, and frequently contain considerable organic matter containing nitrogen.

The Peruvian guano of earlier times was particularly rich in the best forms of nitrogen. The purely phosphatic guanos are rich in phosphoric acid, and are excellent materials; like the iron phosphate, they are not suitable for the manufacture of artificial manures.

Insolubility of Phosphates. — The phosphates mentioned constitute what are called "raw materials," and, with the exception of bone, are not largely used directly, or without further treatment to render the phosphoric acid more soluble, and thus more immediately available to plants. As already stated, the phosphoric acid in them becomes food in proportion to the rapidity of decay, which is influenced both by the character of the material and the fineness of its division. Fine materials, too, permit of a more even distribution, thus bringing more particles of phosphate in contact with the roots of the plants.

CHAPTER VIII.

Artificial and Concentrated Manures; Superphosphates and Potash Salts.

PHOSPHATE of lime is a chemical salt capable of existing in three forms. The first consists of three parts of lime and one part of phosphoric acid; this is the insoluble form, and it exists as such in all natural phosphates. This form, because of the three parts of lime contained in it, is also called "tricalcic," "tribasic," or "three-lime phosphate." The second form consists of two parts of lime and one of phosphoric acid, and is called "dicalcic," "dibasic," or "two-lime phosphate;" it is insoluble in water, but readily soluble to the roots of plants. The third form consists of one part of lime and one of phosphoric acid, and is called "monocalcic," "monobasic," "acid phosphate," or "superphosphate." This form is completely soluble in water, readily distributes itself everywhere in the soil, and is immediately available to plants. A "tetrabasic," or "four-lime phosphate," has been found in basic slag. This form, though insoluble in water, breaks up readily and is more available than the insoluble "tribasic" form.

Superphosphates, or soluble phosphates, are made from the raw materials containing insoluble tricalcic phosphate, by first grinding them to a powder and then mixing them with sulphuric acid, which changes the tricalcic — three-lime — into the monocalcic — one-lime — form, or the insoluble into the soluble form. In this process, two of the three parts of the lime, combined with the phosphoric acid to form the insoluble phosphate, are removed and united to sulphuric acid, forming sulphate of lime, leaving one part of lime combined with phosphoric acid, which is the "monocalcic" or "superphosphate."

A pure superphosphate is, therefore, a mixture of a soluble phosphate, and of sulphate of lime, or gypsum.

Soluble Phosphoric Acid. — Nearly all workable products containing phosphate of lime are capable of being converted into an "acid phosphate" or a "superphosphate." The soluble phosphoric acid thus obtained is a definite compound, and is identical in composition, from whatever source derived.

The term "phosphate" is applied to any material containing, as its chief constituent, phosphoric acid. The term "superphosphate" is applied to any material containing soluble phosphoric acid as its chief constituent.

Thus we have the phosphates already described, which when treated with sulphuric acid are converted into superphosphates, as bone superphosphate, South Carolina rock superphosphate, bone-black superphosphate, bone-ash superphosphate, and Florida rock superphosphate. Care should be taken not to confound the terms "phosphate" and "superphosphate." They are, as we have seen, very different both in composition and character.

Composition of Superphosphates. — Superphosphates differ in their content of phosphoric acid according to the composition and character of the phosphates from which they are made. Those made from organic phosphates, as bone black and bone ash, are richer in soluble phosphoric acid than those made from animal bone or from mineral phosphates; since these materials are of such a character as to enable the manufacturer to add sufficient sulphuric acid to convert all of the phosphate present into a soluble form, and at the same time to secure a dry fine product, which is an important consideration in making superphosphates.

Mineral phosphates, both because of their hardness and of the presence of other minerals which are attacked by the acid, are less easily dissolved, and require more acid in proportion to the phosphate present than those from organic sources. They are also less absorbent, hence it is more difficult to secure good condition when sufficient acid is used to dissolve all the phosphate. In making superphosphates from these materials, less acid is used than is required to completely dissolve the phosphates; and there is, therefore, always present in them more or less of the insoluble phosphoric acid.

In the case of animal bone, too, less sulphuric acid is used than is required to completely dissolve the phosphoric acid; otherwise, a gummy, sticky product would result, due largely to the organic matter in the bone. The insoluble phosphoric acid in bone, bone-black, and bone-ash superphosphates is, however, of greater value than the insoluble in the mineral phosphates, for reasons already given.

In superphosphates, too, there is nearly always present a greater or less amount — depending upon the material —

of the second form of phosphoric acid, the dicalcic, also called "reverted" or "retrograde." This form exists in the greatest amounts in those made from mineral phosphates, which is believed to be due either to the soluble acting upon the insoluble portions, or to the presence of oxide of iron and alumina, which combine with a portion of the soluble phosphoric acid. The soluble goes back to a less soluble form.

In stating the composition of superphosphates, the three forms of phosphoric acid are all recognized. The sum of the soluble and reverted is called the "total available," because these forms are regarded as immediately useful to the plant.

Bone ash and bone black contain on the average sixteen per cent of total available phosphoric acid, practically all soluble; while those from mineral sources usually contain less than fourteen per cent total available, which includes one to three per cent of dicalcic or reverted. These also contain from one to three per cent of insoluble phosphoric acid.

Superphosphates made from animal bone are more variable in their composition than those made from bone black or the mineral phosphates; this being due largely to the variability of the raw materials, chiefly in respect to the content of phosphoric acid. These differ, too, from the others mentioned in containing nitrogen in addition to the phosphoric acid; for this reason they are frequently called "ammoniated superphosphates," or dissolved ammoniated bone.

Advantages of Soluble Phosphoric Acid. — Soluble phosphoric acid, in addition to its direct availability, which is its first advantage, is chiefly valuable because of its ease of self-distribution. When applied to the soil, it is taken up by the water and more generally distributed than is possible by any mechanical means, however fine the substance may be ground. The roots of plants come in constant contact with it wherever they go.

The dicalcic or reverted phosphoric acid is believed to be quite as available to plants, but it remains exactly where it is placed; if the roots are there they can make quite as ready use of it as the soluble. The main difference between the soluble and reverted is that in the former case the phosphoric acid goes to the roots, while in the latter case the roots must go to the phosphoric acid. The same is true of the insoluble, though in a different degree. Here the roots must not only go to the phosphoric acid, but the amount that can be used is measured by the activity of the roots in aiding its solution.

A superphosphate, therefore, is valuable in proportion to the amount of soluble phosphoric acid contained in it; the greater the amount of soluble, and the less the amount of reverted and insoluble, the more valuable. If insoluble or reverted forms are desired — and they are often quite as useful as the soluble — they may be procured from untreated products.

Fixation of Phosphates. — Phosphoric acid, though soluble in water, is not washed from the soil; it is fixed there by combining with the lime and other minerals present. It is believed to assume first, by the appropriation of lime, the dicalcic form, though it is not positively certain that the insoluble tricalcic phosphate is not sometimes formed. It may also combine with iron and alumina and form phosphates; these forms are believed to be less readily taken up by the plant than the dicalcic form.

The time required for this fixation, as well as the form it takes, depends upon the character of the soil; though on soils in a good state of fertility the fixation is quite rapid. On very sandy soils the fixation is sometimes incomplete, because of the absence of lime and iron.

Use of Superphosphates. — Because of the tendency of soluble phosphoric acid to form in time relatively insoluble compounds in the soil, it is often recommended to use a mixture of superphosphate and of animal bone, instead of either alone; the soluble for immediate use, and the less soluble for use at later stages of growth, or for the improvement of fertility.

Superphosphates are never better or more available than when applied; phosphates are probably never less available than at the time applied.

Recent experiments and studies show that fine ground phosphates are very desirable under certain conditions, and their use is gradually growing in favor. This point has reference, however, to the economy of use, which will be discussed in detail in its proper place.

Potash Manures. — Farm crops remove considerable amounts of potash; and since many soils, particularly those composed largely of sand, are not rich in this element, potash becomes a very important constituent of manures.

In the early history of the country, wood-ashes were an important, and practically the only, source of potash for manurial purposes, aside from yard manure and vegetable wastes. At the present time by far the most important source of potash is the Stassfurt mines of Germany. These mines consist of deposits of crude salts, which have doubtless been formed by the evaporation of the water in an inland sea. They have been worked since 1862; and, while enormous quantities have been removed, the extent of the deposits is so great as to appear inexhaustible.

These salts, as mined, contain relatively small percentages of actual potash, and considerable quantities of other salts, some of which are injurious to plants, though a number of the crude salts are used directly upon the land.

The crude products of the mines, shipped and sold in this country, consist chiefly of kainit and sylvinit, and the manufactured products are muriate of potash, sulphate of potash, and sulphate of potash and magnesia.

These salts are all completely soluble in water, and equally available as sources of food to plants. The forms have reference mainly to the effect, good or bad, upon the growth of plants, of the constituents with which the potash is combined, or the other salts with which the potash is associated. Chlorides are believed to be less desirable than the sulphates for certain crops.

Forms of Potash. — The commercial potash salts used are of two distinct chemical forms, — one in which the potash is combined with chlorine to form chloride of potassium, or, as it is more generally called, "muriate of potash;" the other in which the potash is combined with sulphuric acid to form "sulphate of potash."

Kainit. — This is the only crude product that is

largely used directly upon the land. It is composed of a number of salts, chiefly "sodium chloride," or ordinary salt, "magnesium chloride," "magnesium sulphate," and "potassium sulphate."

Although the potash in kainit is in the form of a sulphate, its effect is quite similar to that derived from the use of muriate, because of the large quantities of chlorides contained in it, in combination with magnesia and soda. It is not rich in potash, containing on the average twelve and one-half per cent of actual potash, or potassium oxide.

Sylvinit. — This is a crude salt, similar to kainit in that it contains relatively small amounts of actual potash, though the potash in sylvinit exists both in the form of a sulphate and of a muriate, or chloride. There is, too, in this salt less of the magnesia compounds than in the kainit. Sylvinit is not largely exported to this country. The analyses of the products used here show an average of sixteen per cent actual potash.

Kainit and Sylvinit as Indirect Manures. — These crude salts are valuable as indirect manures in that the salts present, other than the potash, have a solvent effect upon other soil constituents, particularly phosphates; they also aid in many cases in improving the physical character of soils. It is believed, too, that the magnesia contained in them serves as direct food under certain circumstances, though this point is not regarded as of great importance.

The Application of Crude Potash Salts. — In the use of these forms of potash, it is recommended that their application should precede by a considerable time the planting of the crop, in order to avoid danger to the young plant from an excess of magnesia salts, which injure the tender rootlets of plants, and also that the excess of chlorides, which sometimes influence unfavorably the quality of the produce, may be washed from the surface soil by the rains.

In Germany, where the use of these potash compounds has received most careful study, their application is almost invariably made in the fall of the year, or upon the crop preceding the one which is in especial need of potash fertilization. In this country, owing to our heavy spring rains, an early spring application will doubtless answer quite as well in most cases.

Muriate of Potash. — This salt is manufactured from the crude forms, and is the richest in potash of the Stassfurt products. It varies in composition according to the method of manufacture, the commercial products being divided into three grades. The grade most commonly met with upon the markets here contains about fifty per cent actual potash, or potassium oxide. The chief impurity is common salt, or sodium chloride; the lower the content of potash, the higher the content of sodium salts. This form of potash is perhaps more largely used than any other.

Sulphate of Potash. — This form of potash, often called "high-grade sulphate," is regarded as preferable to the "muriate" for many crops, particularly sugar-beets, tobacco, potatoes, and fruit, chiefly because of its more favorable influence on the quality of the produce. It is, however, more expensive than the muriate, and is not so largely used by the manure-makers. Its effect upon yield is not believed to be superior to the muriate. Commercial forms of sulphate of potash contain on the average fifty per cent of actual potash.

Double Sulphate of Potash and Magnesia. — This product is similar to the high-grade sulphate in its effect. It contains, in addition to the sulphate of potash, over thirty per cent of sulphate of magnesia. The potash contained in the product, as usually found, is equivalent to about twenty-six per cent of actual potash, though lower grades are made. These are known under the name of "double manure salts." The magnesia is regarded as of considerable value, particularly in potato manures. The cost of potash in the double sulphate is also greater than in the muriate.

Appearance of Potash Salts. — Although all these products exist in the form of salts, they differ in appearance and character. The sulphates are usually in the form of a fine powder, in color ranging from nearly white to a dirty gray. The muriate is in the form of small, though distinct crystals, varying in color from grayish white to light brown. The kainit is composed of crystals, varying in color from white to dark gray, giving the ground salt a rather pepper-and-salt appearance. Upon standing, all of these salts have a tendency to become hard, though, with the exception of kainit, they are easily pulverized. Kainit often becomes very hard, and requires regrinding in order to make its application possible.

The Uses of Potash Salts. — Although these salts are regarded mainly as sources of potash to the manure manufacturer, their direct use upon the land is increasing rapidly. This is due in large part to the facts that they are of such a character as to make their application, and even distribution, a comparatively easy matter; that the quality of the potash is not improved by the manufacturer; and that on many soils crops respond to liberal applications of potash alone. The crops most benefited are potatoes, white and sweet meadow grasses, clover, and orchard fruits.

CHAPTER IX.

Artificial Manures or Fertilizers; Methods of Buying; Valuation; Formulas.

THE fertilizing materials described in the three preceding chapters are the raw materials, and are the main sources of supply of plant-food to manufacturers and to farmers.

Standard High-Grade Materials. — Such materials as nitrate of soda, sulphate of ammonia, dried blood, bone black, and South Carolina rock superphosphates, and the various potash salts, are called standard products. Different samples of any of these do not vary widely in their composition, and those of the same kind are practically uniform in their action. For instance, any one ton of nitrate of soda contains practically the same amount of nitrogen as any other ton, and the nitrogen is always in the form of a nitrate.

They are standard because they can be depended upon, both in respect to composition and form of the essential element. These are important advantages not possessed by the natural manures or fertilizing materials derived from other sources. They are also called "chemical" or "high grade," because they are in most cases chemical compounds, and because they furnish those particular elements in their most concentrated and active forms.

Incomplete and Complete Fertilizers. — Fertilizing materials may contain but one or two of the essential constituents, nitrogen, phosphoric acid, and potash. Hence the name "incomplete fertilizer" is sometimes applied to them, signifying that they do not serve in all cases to supply the probable needs of the crop.

The fertilizers manufactured from raw materials usually contain all three of these essential constituents; hence they are called "complete fertilizers," signifying that they completely meet the needs of the crop in reference to the number of the constituents that are liable to be lacking.

Methods of Buying. — In buying a fertilizer, that which gives direct value is the fertilizing constituent, nitrogen, phosphoric acid, or potash; hence the transaction is virtually the buying of one or more of these constituents. It is readily seen, therefore, that the more concentrated the product, the less will be the actual cost of the constituent desired.

Again, fertilizers may be bought and used either as "incomplete," — raw materials, — or as "complete," — manufactured products or mixtures, the process of manufacture consisting chiefly in mixing, grinding, and preparing the various materials described. There are certain advantages and disadvantages in both methods of buying. The advantages in the purchase and use of raw materials are : —

1. A better knowledge of the kind and quality of plant-food obtained; that is, these products as a rule possess characteristics which distinguish them from others and from each other, and they are more liable to be uniform in composition than mixtures.

2. The using of one or more of the constituents as may be found to be necessary, thus avoiding the expense of purchasing and applying those not required for the particular crop or soil. The farmer is also enabled to adjust the forms and proportions of the various ingredients to suit what he has found to answer the needs of his soil or crop.

3. A saving in the cost of plant-food, since in their concentrated form the expenses of handling, shipment, bagging, etc., are reduced.

The chief disadvantages in the buying and use of incomplete fertilizers are: —

1. They are not so generally distributed among dealers, and thus not so readily obtained.

2. It is difficult to spread evenly and thinly products of so concentrated a character, particularly the chemical salts, which, unless great care is used, may injure by coming in immediate contact with the roots of plants.

3. The mechanical condition, or degree of fineness, is less perfect than in the manufactured products.

The advantages in the purchase and use of complete manures are: —

1. They are generally distributed, and can be purchased in such amounts and at such times as are convenient.

2. The different materials may be well proportioned, both as to form of the constituents and their relative amount for the various crops.

3. The products are, as a rule, finely ground and well prepared for immediate use.

The chief disadvantages are: —

1. That it is impossible to detect in a mixture whether the materials are what they are claimed to be.

2. That without a true knowledge of what constitutes value, many are led to purchase on the ton basis, without regard to the quantity and quality of the plant-food offered.

Guarantee. — The fact that consumers are unable to determine the value of a mixture from its appearance, and the opportunity thus afforded for disguising the presence of poor forms of plant-food, has led in many states to the enactment of laws which require that all manufacturers shall publish the actual composition of their products, and also state the kind of material from which the constituents have been derived; or, in other words, that they shall guarantee the goods to contain certain amounts and forms of the three plant-food elements, the state exercising a chemical control of the products sold.

By this means, spurious articles are kept from the market, and good manufacturers and farmers are protected, though it is left still to the intelligence of the farmer to determine whether there is a proper relation between the guarantee and selling price.

Interpretation of Guarantees. — The statement of the guarantee is sometimes confusing to purchasers, as different manufacturers use methods which seem to them most desirable. The following examples illustrate this point: —

Guarantee No. 1.

Nitrogen (equivalent to ammonia)	3- 4%
Available Phosphoric Acid (equivalent to bone phos. of lime)	18-22%
Potash (equivalent to sulphate of potash)	10-12%

Guarantee No. 2.

Nitrogen	2.50- 3.25%
Available Phosphoric Acid	8.00-10.00%
Potash (actual)	5.50- 6.50%

ARTIFICIAL MANURES OR FERTILIZERS.   105

The guarantees here given mean practically the same in both cases. In No. 1 the percentages represent the amounts in combination with other elements; while in No. 2 percentages of actual constituents are stated, viz., nitrogen, phosphoric acid, and potassium oxide.

In order to convert the ammonia into its equivalent of nitrogen, the percentage of ammonia may be multiplied by eighty-two per cent, or divided by 1.214; since ammonia is eighty-two per cent nitrogen, and since one part of nitrogen is equal to 1.214 parts of ammonia.

Bone phosphate of lime is forty-six per cent actual phosphoric acid; hence multiplying the bone phosphate by forty-six per cent gives the per cent of actual phosphoric acid. Sulphate of potash is fifty-four per cent, and muriate of potash is sixty-three per cent actual or potassium oxide; hence, to convert the percentages of these forms into their equivalents of actual, they are multiplied by the factors given.

In most raw materials another method of guaranteeing is adopted, because in these the guarantee is simply a statement of their purity. For instance, nitrate of soda is guaranteed as ninety-five per cent pure nitrate; muriate of potash is guaranteed eighty per cent pure muriate, etc., which means that the products are respectively ninety-five and eighty per cent pure, or in other words, that in the case of nitrate five per cent of it is something other than nitrate of soda, and in the case of the muriate twenty per cent of it is something other than muriate of potash.

The factors necessary to use in the conversion of the constituents in their usual form of combination into the actual are shown in the following tabular statement:—

TO CONVERT THE GUARANTEE OF					MULTIPLY BY:
Ammonia	into an equivalent of	nitrogen	0.8235
Nitrogen	" "	"	" ammonia	1.214
Nitrate of soda	" "	"	" nitrogen	16.47
Bone phosphate	" "	"	" phosphoric acid	0.458
Phosphoric acid	" "	"	" bone phosphate	2.183
Muriate of potash	" "	"	" actual potash	0.632
Actual potash	" "	"	" muriate of potash	1.583
Sulphate of potash	" "	"	" actual potash	0.54
Actual potash	" "	"	" sulphate of potash	1.85

The Unit Basis or System.— What is known as the "unit system" of stating the amount of plant-food contained in a fertilizer is sometimes employed; the "unit" means one per cent on the basis of a ton, or twenty pounds. For instance, a "unit" of nitrogen means twenty pounds, and a dried blood guaranteed to contain ten units, means that two hundred pounds of nitrogen is contained in one ton.

This system is largely used by trade journals in stating quotations, particularly for nitrogenous and phosphatic materials. Purchasing on the "unit basis" is the true method, and hence the most satisfactory of any to both the producer and consumer, and should be adopted in all transactions. It means that the consumer secures what he pays for, and the producer is paid for exactly what he delivers.

Purchasers should insist that any material, whether mixed or unmixed, should be accompanied by a guarantee.

Commercial Values.— The commercial value of raw materials is fixed by trade conditions, as supply and demand, usefulness in the arts or manufacture, and market manipulations. The value of these products for fertilizing purposes depends almost entirely upon the constituents contained; hence the actual cost of the constituent is readily determined when the factors, price, and amount contained in a given quantity, are known. The selling price of nitrate of soda, for example, is $48.00 per ton; and as a ton contains on the average three hundred and twenty pounds of nitrogen, the cost or commercial value of nitrogen is, therefore, fifteen cents per pound.

In many States a system of valuation for mixed fertilizers has been adopted, which furnishes a fair method of comparison of different brands. This method assumes that at points of supply a pound of nitrogen in the form of nitrate, of ammonia, or of definite organic compounds, or a pound of available phosphoric acid, or of potash in the form of muriate or sulphate, is practically the same to all manufacturers. A value for each of these constituents derived as already described, when applied to the constituents in the mixture, represents the cost of the elements before they are mixed to form complete fertilizers; and hence the difference between the valuation and selling-price of a brand represents the charges, including profit, for mixing, bagging, shipping, and selling the goods.

This valuation of a brand is commercial, and bears no strict relation to its possible agricultural effect; it simply states that so many pounds of the constituents as are contained in a ton are commercially worth the value given, at point of production. It shows what a given lot or brand of fertilizer is worth as a commodity of trade;

what it costs; and a comparison of the valuation and selling-price of a number in connection with their composition indicates which is the best for the money.

Nitrogen from the same source is worth no more in one brand than in another; the same is true in reference to potash and available phosphoric acid.

Analyses of Fertilizers. — The chemical analysis of a fertilizer should show, as far as possible, both the amount and form of either or all of the three constituents contained; viz., nitrogen, phosphoric acid, and potash. Such a complete statement gives considerable information as to the source and quality of the materials from which the constituents have been derived. For instance, if the analysis shows that three forms of nitrogen are present, that the "total available" phosphoric acid is chiefly soluble in water, that the percentage of insoluble phosphoric acid is low, and that the potash is in the form of sulphate, it is good evidence that stanard high-grade goods have been used.

The analysis cannot, however, give definite and positive information as to the source of organic nitrogen, whether from the best form, dried blood, or from the poorest, leather. Neither is it possible to tell absolutely how much of the insoluble phosphoric acid has been derived from organic or mineral sources, when materials from both sources have been used.

Fertilizer Formulas. — A fertilizer formula indicates the kind and quantity of raw materials to be used to secure certain proportions of the fertilizer constituents. If it be desired to secure a mixture containing four per cent nitrogen, eight and eight-tenths per cent available phosphoric acid, and ten per cent of actual potash, the following materials would furnish it, assuming an average analysis for them: —

Formula No. 1.

MATERIALS.	AMOUNT.	Nitrogen.	CONTAINING POUNDS OF Phosphoric Acid.	Potash.
Nitrate of Soda	500 lbs.	80		
Bone-black Superphosphate	1,100		176	
Muriate of Potash	400			200
Total	2,000	80	176	200
Guaranteed Analysis		4%	8.8%	10%

A mixture containing two and one-half per cent of nitrogen, eight per cent of available phosphoric acid, and two per cent of potash may be made from the following materials: —

Formula No. 2.

MATERIALS.	AMOUNT.	Nitrogen.	CONTAINING POUNDS OF Phosphoric Acid.	Potash.
Nitrate of Soda	150 lbs.	24		
Dissolved Bone	1,300	26	160	
Muriate of Potash	80			40
Land Plaster	470			
Total	2,000	50	160	40
Guaranteed Analysis		2.5%	8%	2%

No. 1 is a high-grade product, both in respect to quality of plant-food and concentration; while No. 2 is high-grade only in respect to quality. In order to get the plant-food distributed throughout the ton, it is necessary to add what is called a "make-weight" or diluent.

High-grade mixtures cannot be made from low-grade materials. Low-grade mixtures cannot be made from high-grade materials without adding "make-weight." The advantages of high-grade products are concentration and high quality of plant-food.

Special Formulas.— Frequently a large number of different formulas or brands are placed upon the market by the same manufacturers. The claimed purpose in the multiplication of brands is to meet the various demands of the consumer, as well as to furnish special preparations which shall provide a large proportion of that constituent which is believed to be of special service to the particular crop.

For instance, special potato manures contain a much larger proportion of potash than those intended for general purposes. Formula No. 1 may be regarded as a special potato manure, while No. 2 may be regarded as a general formula. It must be remembered that the amount of plant-food applied frequently exercises a greater influence than mere proportion of the elements contained in it. The multiplication of brands is seldom of advantage to the consumer.

The Use of Fertilizers.— To use fertilizers to the greatest advantage it is requisite that a great many points should be carefully studied. The character of the manure itself; the soil and previous treatment, both in reference to manuring and cropping; the climate; the character of the crop to be grown, and the object of its growth,— are, perhaps, the chief factors to be taken into consideration.

It has already been pointed out that the manurial constituents exist in various degrees of availability, from complete solubility in water to insolubility except in strong acids; the character of both the soluble and the insoluble determines its usefulness to the plant. A correct knowledge of the action of these is, therefore, of the first importance in order to economically use them.

Nitrogenous Manures. — In reference to nitrogenous manures, it may be stated, that, because nitrogen in the form of a nitrate is immediately available, and because it is freely movable, and is not retained by the soil, nitrate should not be applied in any considerable amount before the plant is growing and ready to use it. While in the case of nitrogen in the form of ammonia, though it is completely soluble in water, it is absorbed by the soil, and requires an appreciable time to change into the form of a nitrate, and may, therefore, be applied without risk of loss a short time before it is likely to be used.

On the other hand, nitrogen in organic forms shows a wide range of availability, the readily soluble blood ranking with the nitrate and ammonia; while leather, wool waste, shoddy, and like products, which decay very slowly, should be applied a considerable time before the nitrogen in them is required.

Phosphatic Manures. — In the case of phosphatic manures, the soluble forms, or superphosphates, should be applied but a short time before the plant requires the food, since their tendency in the soil is to revert to their original and insoluble forms. Coarse bone, ground mineral phosphates, and products of like character, decay slowly, and should be applied a long time before they are likely to be used.

Potash Manures. — The potash manures from the Stassfurt mines are readily soluble; they should, however, be applied some time before they are required, in order to secure their complete distribution in the soil. On soils of a heavy character a fall application is recommended.

In order to attain the best results from mixed fertilizers, great care should be given to the proper adjustment of the various kinds and forms of the materials used.

Kind of Soils. — Crops grown upon soils poor in decaying vegetable matter are, as a rule, benefited by nitrogen manuring, while those upon soils rich in this substance are more benefited by phosphates and potash. Upon heavy soils phosphates are likely to be more beneficial than nitrogen, while the reverse is the case on light, dry soils. All sandy soils are, as a rule, deficient in potash, while clayey soils contain this element in larger quantities.

Different Methods of Growth. — The difference in crops in reference to their capacity for acquiring food must also guide in the application of manures. Crops that have deep roots, and grow throughout a long season, are able to acquire their necessary food where those of shallow roots and short seasons of growth would suffer hunger.

Crops of the same class, too, resemble each other to some extent in their capacity for acquiring food. The grasses, for example, do not possess a strong power of assimilating nitrogen; root crops possess a small capacity for acquiring and utilizing phosphoric acid; while leguminous plants are unable to readily assimilate potash; hence these crops are, in the order given, most benefited by nitrogen, phosphoric acid, and potash.

CHAPTER X.

The Rotation of Crops.

THE aim of the farmer, as well as those engaged in other industrial pursuits, is to derive the greatest possible return both for his labor and money invested. The selection of definite lines of farming, or the growth of crops profitable for his conditions, becomes, then, of great importance.

The Demand for Special Crops. — In the earlier history of the country, selection was practically limited to the staple crops of grain and hay. As the country developed and increased in wealth, larger demands were made for fruits, vegetables, and such special products as were in former times regarded as luxuries, and the production of which was confined to the areas of gardens and yards.

At the present time, therefore, particularly in the Eastern States, general farming is the exception rather than the rule, and special farming is more profitable. The raising of hay, grain, vegetables, and fruits, and dairy products, now forms distinct lines. The adoption of either or any of these depends upon a variety of circumstances, though chiefly upon the following: the conditions of soil and climatic influences; the location of the farm in respect to markets; and the probable profit.

In any or all of these lines, however, certain groups of crops may be more profitable than others. This is because it has been found to be more desirable in the long run to have a variety, one following the other in a definite rotation. In the cotton and sugar producing States of the South, and the wheat growing States of the Northwest, rotation is least practised, while in the Eastern States and the Central West, rotations are the rule.

The practice of growing different crops in rotation, while largely a matter of conditions, does possess certain advantages, — based upon scientific principles, as having reference to the character of growth and feeding capacities of plants, and upon business principles, as having reference to a better division of labor and a more certain income.

The Advantages of Rotations. — The advantages of rotations may be stated as follows: —

1. The feeding capacities of plants differ, certain of them requiring more of one particular element than of another; certain are surface feeders, and others send their roots deep into the subsoil. The growth of a variety of plants with different capacities, therefore, prolongs the period of profitable culture, or retards soil exhaustion.

2. The growing of but one crop leaves the soil bare at certain seasons of the year, while a variety permits of a continuous growth and covering of the soil. Soils suffer loss when lying idle; they are improved by the growth of crops.

3. The continuous growth of one crop renders it more liable to insect attack, and to the development of dis-

eases that cause rot and blight. Crops lose vigor by being grown year after year, and thus are less able to withstand these attacks; besides, a change of crops deprives the pests of their particular kind of food, causing them to disappear or perish.

4. Certain crops derive their nitrogen, phosphoric acid, and potash entirely from the soil; the cereals grown for their grain, which is usually sold, belong to this class. Certain others, the clovers, peas, and beans, derive their nitrogen from the atmosphere; their removal does not decrease the store of nitrogen in the soil. A rotation of crops, including the latter, therefore, lessens the necessity for nitrogenous manuring.

5. A rotation of crops distributes labor throughout the season, thus giving continuous work for men and horses. In farming districts it is difficult to procure labor for short periods, while horses have to be kept throughout the year.

6. A variety of crops marketed at different periods, permits a steady and regular income to the farmer, which enables him to do business on a smaller capital; wages can be paid when due, and his supplies of seeds, fertilizers, implements, and tools can be bought in the lowest market for cash.

It has already been stated that the adoption of what are now "systems of crop rotation" was largely a matter of growth, due to circumstances, and was not in the beginning based upon scientific principles. Science, however, furnishes the reasons why rotations are useful, and why certain rotations are more useful than others.

The Need of Rotations. — The need of rotation as

a means of maintaining fertility was apparent in early times, when the manures were confined to the natural wastes of the farm, and when the growth of livestock and production of dairy products were industries of but little importance. It is less apparent now, when the materials furnishing available plant-food elements or artificial manures are so abundant and cheap. Formerly, the proportion of active soil constituents was almost entirely dependent upon the natural forces that were brought to bear upon the dormant constituents to convert them into activity; under the conditions that exist now, it is frequently more economical to purchase the active constituents and apply them to the soil; in other words, to supplement natural forces by artificial.

Bare Fallow. — In the older systems of rotation, it was customary to allow the land to lie bare, or "fallow," once in two or three years, in order that the natural agencies, sun, air, and water, might have free access, cause a more rapid decay of the soil particles, and make it more fertile, a practice extending the period of profitable cropping without manure. Sometimes the fields were left entirely to themselves, while in others they were frequently plowed or stirred in order to hasten the decay.

Fallow Crops. — Following this method came "crop-fallowing," which is still practised; that is, instead of allowing the land to remain idle after a grain crop has been removed, a cultivable crop, as turnips or roots, is planted, or a catch crop, as clover, is seeded, the cultivation of the one assisting in the decay of vegetable and mineral matter, thus improving for a subsequent grain crop; while the other, because of its different method of growth and greater power of acquiring food, assists in renovating and improving the soil.

Rotations to be Adopted. — The rules which govern the adoption of systems of rotation, under the conditions that now exist, are general and flexible, rather than specific and fixed. To grow the crops that pay the greatest profit per acre should be the aim, and rotations should be modified in such a way that the least profitable crops should contribute as much as possible to the development of the most profitable. The character of soil, climate, availability of farm-labor, location, markets, — all have an influence in determining what the most profitable crop may be.

For instance, hay may be high in price in a given locality: the soil is dry and sandy; hay burns on the ground; the yield is light, and it does not pay to raise it, even at high prices. In another locality sweet potatoes may bring three dollars per barrel: the land is a clay, cold and heavy; it is suitable for hay, not sweet potatoes. Reverse the order, and both may be profitable crops. On the light, sandy land the rotations adopted should be such as contribute to the best development of the sweet potatoes, and on the heavy clay, such as aid in preparing the soil to produce the largest hay crop.

The climate, in the same manner, places a limit upon the production of certain crops. A short, cool season is not favorable for the corn crop; it will not mature: hence corn should not be included in a rotation under such conditions. In many cases farms do not pay because their owners have not studied their conditions in reference to paying crops, and adapted themselves to them.

Rotation Courses. — The number of years intervening between the growth of crops in regular order is termed a "course." A rotation course may range from the simple two-year to the more complex six or eight year course, though the four-year course is generally adopted. The poorer the land the shorter the course; and the reverse, the better the land the longer the course, — are principles now well established.

Taking the number of crops and periods of rotation possible, it is evident that the number of possible courses is too large to admit of definite description or comment. A few examples only are given and discussed, in order to more clearly illustrate the principles already pointed out.

Examples of Rotation Courses. — These are adapted to what is termed "arable farming," where the live stock is only sufficient to provide labor and the necessities of the family.

	First Year.	Second Year.	Third Year.	Fourth Year.
1.	Corn.	Oats.	Wheat.	Clover.
2.	Corn.	Wheat.	Clover.	
3.	Corn.	Potatoes.	Wheat.	Clover.
4.	Corn.	Potatoes.	Clover.	
5.	Potatoes.	Wheat.	Clover.	

No. 1 is defective for two reasons: first, because uncultivated crops similar in character and capacity of obtaining plant-food succeed each other; and second, because the oats preceding the wheat prevents a proper cultivation of the soil and preparation of the seed-bed for wheat. This rotation is widely used, mainly because it is economical of labor. Until very recently the custom was to plant the corn on clover-sod, follow with oats without manure, and then lime and manure for wheat.

This custom is now rapidly changing, and for the better; viz., to manure the corn, and to provide artificial fertilizers for the oats and wheat. The latter method is more reasonable, since it permits of the removal of the manure from the yard to the field during the leisure of winter and spring, and the increased profit from its use is received in the year of its application. It is more economical of labor and capital. In all the rotations where clover follows wheat, it is usually seeded in the growing wheat in early spring.

No. 2 is particularly adapted to light lands, as it admits of a more frequent repetition of the renovating clover crop. It is objectionable, however, where the seasons are short, since to wait for seeding the wheat until after the corn is fit to harvest does not allow it to make sufficient top to withstand the winter well; besides, the early removal of the corn is very laborious and expensive.

No. 3 is a typical rotation, since the crops of cereals are separated by a root or clover crop. This rotation corresponds to the Norfolk system, so widely adopted in England; viz., turnips, barley, clover, wheat. In many sections of this country the potatoes are the best paying crop. The corn is planted on a clover-sod, and yard-manure liberally applied; the corn, being a gross feeder, utilizes sufficient food for its normal growth from the partial decay of the manure, roots, and stubble, and the cultivation of the corn puts the land in excellent tilth for the potatoes. Artificial manures are mainly used for this crop, frequently in large amounts, the residues from which guarantee maximum crops of both wheat and clover.

In No. 4 wheat is dispensed with, and in No. 5, corn. Both are excellent where potatoes or root crops can be grown to advantage; and, if the land is naturally rich, the frequent tillage, and use of clover crops, provide an abundance of available food for maximum crops, provided the second crop of clover is not removed.

No. 5 may be reduced to a two-year rotation by plowing the clover in spring before removing any crop. In these rotations barley may be substituted for oats, rye for wheat, and sweet potatoes or tomatoes for potatoes, without interfering with the usefulness of the rotation.

Rotations on heavy land, where hay is an important crop, differ mainly from those already mentioned in having a larger number of crops. Timothy is seeded with the wheat in addition to clover in the spring. The first year after wheat, a mixed hay crop is gathered, which becomes almost pure timothy in the next season, and purer still in that following; hay is cut two or three years or longer, as the strength and character of soil permit. Cropping in this way is, however, exhaustive, and requires careful manuring. These rotations have reference to what is termed "extensive practice," and do not provide for the manuring of each crop, though it does not follow that it cannot be made "intensive."

Rotations in Market-gardening and on Dairy Farms. — In market-gardening and dairy-farming, manures are relied on to a greater extent, and less attention is given to strict rotations. These lines of farming are more on the "intensive plan," the areas are limited, the cropping constant, the manuring liberal, and the crops as large as conditions of climate and season will permit.

The market-gardener, as soon as one crop is removed, plants another, keeping the land constantly occupied. He does not depend upon natural fertility, but forces growth by the abundant supply of natural and artificial manures. The rotation practised is governed by the conditions which control the kind of crops he can grow to advantage, rather than by considerations of soil fertility.

In dairying, the object is to provide a continuous supply of food; hence the rotation adopted is the one which will best meet this requirement.

CHAPTER XI.

The Selection of Seed; Farm Crops and Their Classification; Cereals; Grasses; Pastures; Roots; Tubers; and Market-garden Crops.

Selection of Seed. — The kind of seed used exercises an important influence upon the yield and quality of the crop, and also saves the farmer losses due to a poor stand. The larger the proportion of living seed true to kind, the greater the chances of a perfect stand and a normal and healthy growth of crop. In the case of the larger seed, as the cereals, it is not so difficult to determine quality as in the case of certain grasses and garden seeds; here a careful examination and testing are required.

Good Seed. — The term "good seed" implies that any given lot should show a large proportion of mature seed, true to kind and variety, and a small proportion of impurities and adulterants. "Mature seed" are those that have fully ripened, and are capable of performing well all of the functions of germination; that is, they are capable of using the food stored up in them, and developing vigorous and healthy young plants. "Immature seed" are those that have not fully developed or ripened, and can only partially perform the functions of germination. The young plant lacks strength and vigor.

Impurities. — Impurities include all foreign matter,

both injurious and harmless, that may be present in the seed purchased, as seeds not genuine, dirt, dust, weed-seed, chaff, and diseased seed. The presence of weed-seed perhaps results on the whole in the greatest loss and annoyance, — in the first place, the loss of return from land taken up by the weeds, hence a reduction in crop; and second, the difficulty and expense of eradicating the weeds when well established.

Adulteration. — Adulteration of seed includes, first, the substitution of cheaper seed for the more valuable, which is frequently practised in the case of mixed seeds that resemble each other, and second, the removal of the evidences of age or disease. Seed that are musty or dark are sometimes sweetened and brightened by bleaching with fumes of sulphur.

Quality of Seed. — The quality of genuine seed is influenced by age, size, weight, and smell. Old seed are less likely to germinate than new; the loss of vitality is gradual, though more rapid in unripe than in well-ripened seed; the larger and heavier seeds also die more slowly than the smaller and lighter ones.

The seed of the cereals and grasses lose germinating power and vigor rapidly after the first year; though alive, they germinate and grow slowly, thus causing a loss of time at the beginning of the season, and the slow growth of the plant at its most tender stage increases the tendency to disease and insect attack.

Seed may also be killed by a too rapid or too complete removal of water from them; hence artificial drying, if improperly conducted, that is, if too great heat is used, may result in the death of the germ.

If a crop from which seed is to be gathered is stored before thoroughly dry in a damp place, it is liable to become hot, which destroys in a great degree the germinating power of the seed. Crops from which seed is to be secured should be carefully dried, and stored in a dry place.

Change of Seed. — The improved varieties of farm crops of the same kind have been developed by the careful selection of the best seed of these crops grown under the most favorable conditions of climate, season, soil, and management. The natural tendency of the plant, even under favorable conditions, is to go back to its original and inferior state; hence, when the conditions of growth are unfavorable, this tendency is increased. A change of climate, a season too cool or too hot, too dry or too wet, a poor soil, lack of care in cultivation, — all aid in increasing this backward tendency; the conditions are not perfect, and the seed, as it is commonly expressed, "runs out," and a change becomes necessary.

In making the change, seed should never be taken from good to poorer conditions, but rather from poor to good; that is, the seed from crops grown under good conditions of climate, soil, and management will not retain their character so well when grown under conditions poorer in these respects, while the seed from crops which flourish well under poor conditions are likely to not only retain their character, but improve when changed to good conditions.

It is also true that seed from crops that do well in rigorous climates are more likely to improve when brought under more favorable conditions in this respect than when those that do well in a warm climate are brought into a colder climate. In other words, in changing seed, particularly of the cereals, they should be secured from the North rather than from the South. These are, however, general suggestions, to be used as guides rather than as specific and definite rules.

Seed-Testing. — The number of pure seed and the germinating power are the two factors which determine the number of plants that may be obtained from a given quantity, rather than the bushels of seed sown. Seed-testing includes, therefore, a test of purity and of germinating power.

Purity. — The purity may be tested by taking a definite weight to represent the product, and separating the foreign matter, either by hand or by means of a sieve, then weighing the remainder of pure seed. For example:

> Total weight taken . 100 grains or grams.
> Weight of pure seed . 95 " "
> Weight of impurities . 5 " "

The amount of pure seed is, therefore, 95 per cent, while the amount of impurities is 5 per cent.

It is better in stating the impurity, consisting of foreign seeds, weed-seeds, etc., to use the number of seed instead of their weight, as it gives a better idea of the possible damage from its seeding.

Germinating Power. — In testing the germinating power, only the seed true to kind are tested; hence, a high germinating power is not in itself sufficient evidence of quality. It must be accompanied by a statement as to purity; for instance, if the germinating power is ninety per cent, and the purity only twenty-five per cent, the quality of the seed is low, since out of one hundred pounds only twenty-two and five-tenths pounds consist of pure germinating seed. Good seed shows a high percentage of both purity and germinating power.

To test germination, two lots of at least one hundred seeds each are selected, and placed under conditions favorable for germination; viz., moisture, warmth, and air. A box containing a thin layer of fine soil, kept well moistened, and in a warm place, answers the purpose nicely. The chief precautions to observe are to keep the material moist and the temperature between 80° to 90° F.

Plants are classified by botanists into families or natural orders; by farmers into groups, made distinct by methods of rotation or other local causes.

Botanical Classification. — This is a useful guide to the farmer in indicating habits of growth, as well as methods of manuring and management, since plants of common origin or ancestry, though differing in outward form, are quite likely to be benefited by the same kinds of food, to be subject to the same class of diseases, and to be attacked by the same kind of insects. The different families also include a number of plants not useful as farm or garden crops, though all possess certain characteristics in common.

The Grass Family. — The first natural order in point of usefulness is the grass family. This includes the cereals, wheat, rye, oats, barley, corn, and rice, valuable chiefly for their seed or grain; and the grasses, valuable as hay and pastures, of which timothy, orchard grass, rye grass, blue grass, and red-top, are prominent examples.

The Family which includes the Potato, valuable

for starchy food, the tomato and egg-plant, useful for their fruit, and tobacco, a plant of great commercial value, is also very important. Many plants of this family are poisonous, of which "henbane" and deadly "nightshade" are examples; in fact, the fruit and vine of the potato contain poisonous principles. In point of food value the potato stands next to the cereals.

The Sweet Potato belongs to the morning-glory family, and is the chief food-plant belonging to this order.

The Legume or Clover Family includes a very large number of plants, as herbs, shrubs, and trees. A distinguishing feature of this order is the formation of seed in a pod or legume. Those in which the seed or grain is used as food, as peas, beans, and lentils, are called "pulse;" hence this name has been extended to all the food plants of this order. Leguminous crops are called "pulse crops."

The other plants of this family, useful as hay, green forage, or pasture, are the various clovers, white or Dutch clover, red clover, alsike or Swedish clover, and crimson or scarlet clover, also lucerne or alfalfa, vetches, lupins, serradella, and sanfoin. These plants are among the most valuable of our forage crops. They have strong foraging powers for mineral constituents, and are also able to secure the nitrogen necessary for their growth from the air; thus they enrich, rather than impoverish, the soil of the most important element, — nitrogen.

The Turnip Family includes among the edible plants, turnips, the various varieties of cabbage, as cauliflower, kale, kohlrabi, mustard, radish, horseradish, and watercress, and the forage plants, swedes and rape. This order also includes a large number of weeds, of which shepherd's-purse, charlock, and wild radish are prominent examples.

The Beet Family includes the food-plants, garden-beets, sugar-beets, and spinach, and the mangel-wurzel, an important fodder plant. The sugar-beet is of great importance in Germany and France as a commercial source of sugar, while the mangel-wurzel is extensively raised as a fodder crop.

The Melon Family is extensive, though it does not include any strictly farm crops. Cucumbers, melons, pumpkins, and gourds are prominent examples of this order: "cucurbs" is a term also applied to this group of plants.

The Carrot Family includes carrots, parsnips, parsley, and celery, while lettuce belongs to the dandelion family, and the onion and asparagus to the lily family.

The Rose Family is an important natural order. It includes herbs, shrubs, and trees, to which belong the most important of our fruits. One type is represented by the plum, peach, cherry, apricot, nectarine, and almond; another, by the raspberry, blackberry, and dewberry; and still another, by the apple and pear. The bush-fruits, gooseberry and currant, are both members of another distinct natural order.

Agricultural Classification. — While the grouping of plants by the method described is useful, an agricultural classification, which groups the various crops according to their similarity of growth, management, and treatment, is also convenient to the farmer; and the following method permits of a logical discussion of the principles involved in their growth: —

1. Cereal crops: Wheat, rye, oats, barley, and corn.
2. Forage crops: Grasses and clovers for forage hay and pasture.
3. Root crops: Turnips, swedes, carrots, and mangels.
4. Tuber crops: White potato and sweet potato.
5. Miscellaneous crops: Market-garden and fruit crops.

Cereal Crops. — The chief object in the growth of cereals is to obtain the grain or seed. They are all annuals, though their natural period, or time of growth, differs, certain of them requiring a longer time for their growth and maturity than others. In the case of wheat, rye, oats, and barley, the natural period of vegetation has been changed by careful selection and breeding, so that we have both winter and spring varieties, the former seeded in the fall, and the latter in the spring. The winter varieties of wheat and rye, and the spring varieties of oats and barley, are more generally grown. Indian corn, or maize, is seeded in spring only.

The Root System. — In the cereals, the roots branch just below the surface, and each shoot produces feeding-roots, which distribute themselves in every direction to gather food and directly nourish the plant. The roots of the cereals, though they are regarded as shallow feeders, also penetrate to considerable depths, — thirty-six inches or more, — the depth corresponding to some extent with the period of growth, winter wheat and rye showing the deepest roots, and oats and barley, seeded in spring, the shallowest.

The character of the soil also exercises an influence in this respect. The deeper the root, the more food is acquired; and the power to resist drouth and other unfavorable conditions is proportionately increased. A soil too dense and hard prevents the penetration and development of the root system. The root formation of the winter cereals is encouraged by close contact with moist earth, hence a soil of a compact nature is desirable for their normal growth; loose, shifting sands are unfavorable, moist clay loams are favorable.

Method of Feeding. — The general tendency of the cereals is to absorb food from lower layers of the soil as the plant grows older; that is, the roots near the plant die; and only the fibrous roots at a distance and in the lower layers of soil possess the power of absorbing food: hence, to ensure maximum and continuous growth throughout the whole period of life, the entire surface soil must be enriched.

Power of Acquiring Food. — The cereals are able to acquire food from the insoluble phosphate and potash compounds of the soil in a greater degree than root crops. They are on this account called "voracious feeders." Oats and rye possess this characteristic more largely than wheat or corn. Where the climate is suitable, oats and rye will — other conditions being equal — thrive proportionately better on poor soil than wheat or corn.

These crops are unable to feed to any extent upon the insoluble nitrogen of the soil; they absorb the nitrogen necessary chiefly in the form of nitrates. This form of nitrogen must, therefore, be directly applied, or the soil must have been previously supplied with nitrogenous materials that decay readily.

On soils well supplied with mineral constituents, wheat, oats, and rye — both because they are uncultivated crops, and because their greatest development is in the early summer, before the conditions are favorable for the rapid change of organic nitrogen into nitrates — are more benefited by a direct application of nitrates than corn, which, besides being a cultivated crop, makes its greatest growth in late summer, when the decay and consequent nitrification of the organic matter is most active.

Soil Exhaustion. — The cereals are exhaustive crops, because the food constituents gathered are largely transferred from the stem and leaf, and concentrated in the grain, which is sold from the farm.

Forage Crops; Grasses. — Nearly all varieties of grasses are perennial, though the length of life depends upon the method of cropping and character of soil. Where the grass is allowed to seed, it dies quicker than when it is pastured or cut before maturity, because the depth of root is measured to some extent by the length of top. On poor, dry soils, also, the life is shorter than upon moist soils of fair fertility.

Methods of Growth. — The grasses send out their fibrous roots into the surface soil in the same manner as the cereals, though they differ from the cereals in forming each year a set of buds just below the surface of the ground, which become active in the late summer, and develop new shoots and roots; as this budding ceases, the plants die. Those which produce the greatest number of branches, and continue the process for a succession of years, are the most valuable permanent grasses; those that form their branches in compact tufts have less hold upon the ground, and are more liable to be uprooted by animals, and to be destroyed by unfavorable conditions of soil and season, than those which possess a loose-branching system. A mixture of the tuft and loose-growing grasses is, as a rule, better than either singly.

Food Requirements. — The chief object in the growth of grasses is to obtain the nutrition contained in leaf and stem in the form of pasture, forage, or hay. Nitrogen, which promotes this form of growth, is an important constituent, and it is essential to provide a liberal supply, either directly, as nitrate of soda, or in organic forms, which decay more or less rapidly.

The grasses resemble the cereals in their power of acquiring mineral food; hence clay soils, which are rich in the minerals, are naturally well adapted for their growth. Except when seed is grown, or hay is sold, the grasses are not regarded as exhaustive crops.

Clovers. — The clovers are grouped with the grasses because usually grown for the same purpose, — for pasture, forage, or hay. They, however, belong to a family of plants which possesses characteristics very different from the grasses, both in reference to method of growth and composition of product. The varieties usually grown are : —

Scarlet or Crimson Clover,	an annual.
Red or Broad Clover,	a biennial.
Alsike or Swedish Clover,	a triennial.
White or Dutch Clover,	a perennial.

Of these the red and white are more extensively grown than the others.

Methods of Growth. — The red, alsike, and crimson clovers all possess a tap-root which penetrates downward to considerable depths; and as it descends it throws out root-fibres into the different layers of soil; these gather the food which passes through the tap-root into the branches. The branches are formed from buds, which depend for their food upon the tap-root and its feeders, the fibrous roots. From spring seedings of clover the buds begin to develop in the late summer, lie dormant through the winter, branch forth in the spring, and develop into the mature plant.

With the exception of the annual crimson clover, the process continues two, three, or more years, as the case may be. Crimson clover is usually seeded in the late summer or fall; it develops the buds in the spring, and dies after maturing the plant.

In white clover the stem creeps along just under the surface, throwing out roots at frequent intervals, which penetrate the soil and gather food that is carried directly into the stem, and by it to the branches.

Power of Acquiring Food. — The clovers readily acquire food from the mineral constituents of the soil, and differ from the cereals in being able to acquire their nitrogen from the air; hence on most soils the application of this element is not so essential for their growth. The clovers demand an abundance of potash and lime.

Soil Improvers. — The clovers are not exhaustive crops, but rather soil improvers. The nitrogen gathered and stored as organic substance in roots and stubble enriches the soil in humus and nitrogen, while the method of growth, viz., the formation of large tap-roots, which penetrate deeply, materially improves the physical character of soils.

Root Crops. — Turnips, carrots, parsnips, beets, and mangels are biennials. The first year or period of growth is the storage or vegetative period, and the second, the period of seed-making.

Root crops are all provided with tap or storage roots, and hence they flourish well only in deep, mellow soils. They are supplied with absorbing roots, which spring mainly from the lower end, spread into the soil, and gather the food. The chief supply of food is needed late in the season, when the formation of tap-root is most rapid.

These plants cannot make ready use of the insoluble mineral constituents of the soil; hence, in order to insure full crops, they must be liberally supplied with available food. Of the three classes of fertilizing materials, the phosphates are especially suitable for turnips, while the slower-growing beets and carrots require an abundance of nitrogen in quickly available forms.

Tuber Crops. — The white potato tuber is not a root, but an enlarged underground stem. The true feeding-roots are produced by the underground portion of the main stem. The extensive growth of the plant underground requires that the soil shall be loose and open, in order to permit the free entrance and circulation of both air and water. Potatoes, like root crops, do not possess strong foraging powers. The food must be in a soluble and available condition, in order to insure maximum production. Where soils are suitable, potash seems to be the ingredient especially useful in the manures applied, as it is a potash-demanding plant.

The sweet-potato tuber is an enlarged root, and not a stem, as is the case with the white potato. The plant is especially adapted to warm, dry soils; and, while it thrives well on soils too poor and sandy for the white potato, it must be well supplied with the mineral constituents, particularly potash.

Market-Garden Crops. — These include a large number which are distinguished not so much by their place or method of growth as the object of their production; viz., earliness and succulence, rather than maturity of crop. Lettuce, beets, spinach, radishes, onions, cabbage, turnips, celery, asparagus, tomatoes, egg-plant, cucumbers, melons, peas, beans, sweet corn, and many others, are included under market-garden crops.

To accomplish the two particular objects of their growth requires a deep, warm soil, well supplied with vegetable matter and with available forms of plant-food. Since nitrogen is the element which encourages and stimulates leaf and stem growth, its application, particularly in the form of nitrates which are immediately available, is especially useful for all of these crops; and though peas and beans belong to the legume family, they are materially benefited in their early growth by a supply of soil nitrogen.

Fruit Crops. — These differ from other crops in that there must be a longer season of preparation, in which the growth shall be so directed as to prepare the tree for the proper development of a different kind of product; namely, the fruit. The fruit, too, differs very materially in its character from that of ordinary farm crops, in that its growth and development require a whole season; it is necessary that there shall be a constant transfer of the nutrition from the tree to the fruit throughout the growing-season. The growth for each succeeding year of both tree and fruit is, too, dependent upon the nutrition stored up in the bud and branches, as well as that which may be derived directly from the soil.

Soils that are naturally well adapted for fruit-growing must possess a good physical character; that is, they must be sufficiently open and porous to permit the penetration and growth of the roots, as well as a free movement of air and water, and they must contain nitrogen and the mineral constituents, lime, potash, and phosphoric acid, in considerable amounts.

The first object should be to secure a good tree, though it is not wood growth alone that should be kept in mind, but the *kind* of growth as well; that is, it must not only be vigorous, but well matured. Well-grown trees sometimes produce poor fruit, but poor trees never produce good fruit.

Manures for Fruits. — In the first place, there should be sufficient nitrogen to provide for an abundant leaf growth *early* in the season, since the tree and fruit are dependent for food both upon the leaves and upon the roots. There should be an abundance of potash and phosphoric acid and lime, in order to secure a normal and solid growth of stem and branch, which carry the fruit-spurs, and the food necessary for their first development in the spring, as well as to provide for the proper growth and ripening of the fruit.

CHAPTER XII.

The Growth of Animals; The Constituents of Animals and Animal Food; The Character and Composition of Fodders and Feeds.

IN our study of the growth of plants, it was shown that, with the exception of the food stored in the seed, the plant was built up of single chemical elements, and that these were derived from sources outside of itself, viz., the atmosphere and soil, and formed by the living plant into organized plant substances. The growing of a plant, therefore, is a constructive process; elements that exist separately are gathered from different sources, and combined and fixed in special forms.

The growth of the animal is more complicated. It is built up and nourished by the consumption of substances ready formed in plants, or which have been derived from them. It is a double process, — first, a disorganizing, or tearing apart of these substances formed in the plant; and second, a building or forming process in which they are brought together again, and fixed in the form of flesh and bone.

Composition of the Animal Body. — The animal body, therefore, is composed of substances or elements common to the plants from which it was directly or indirectly derived. It may be divided, first, into two classes of products, water and dry substance.

Water. — This is an important constituent. It is essential to the proper distribution of the nourishing fluids throughout the system, and usually constitutes more than one-half of the total weight of the live animal. As is the case with plants, it is contained in the greatest amounts in the young or immature, and decreases as growth proceeds, and maturity is reached. It is possible to remove it from animal substances without entirely destroying their form, thus differing from the constituents that constitute the dry matter; these cannot be removed without destroying the character of the substances themselves. Here, too, is a striking analogy between plant and animal.

Dry Matter. — This may also be divided into two general classes, — first, that which is organic or volatile, or that portion capable of being destroyed, or converted into gaseous substances, by means of heat; and second, ash or inorganic, mineral, or non-volatile, or that portion which cannot be destroyed or dispelled by means of heat.

Organic Substances. — These are usually divided into two general classes, — first, non-nitrogenous, or those free from nitrogen, consisting of carbon, hydrogen, and oxygen only; and second, nitrogenous, those containing nitrogen in addition to the carbon, hydrogen, and oxygen.

Non-Nitrogenous Substances. — The chief of these is fat, a substance extremely rich in the element carbon, and a very important constituent of food. It is found distributed throughout the various organs of the animal body, though mainly enclosed in cells on the kidneys and between the muscular fibres. The fat contained in the various kinds of animals, which is a whitish, oily substance, differs but little in appearance, though somewhat in composition.

Nitrogenous Substances. — These consist chiefly of carbon, hydrogen, oxygen, and nitrogen, though phosphorus and sulphur are always present in small amounts. This class may be subdivided into three groups: 1. Albuminoids; 2. Gelatinoids; 3. Horny Matter.

Albuminoids. — These are the most important, because animal life is dependent chiefly upon them and the organs composed of them, and because they furnish the material out of which the other groups are formed. They are found in various forms in the body, the chief of which are albumen, fibrin, and casein. These, while differing widely in appearance, agree in their chemical composition, in that each in a pure, dry state contains about sixteen per cent of nitrogen, and from one to one and five-tenths per cent of sulphur; albumen is represented by the white of egg, fibrin by the white solid remaining after the red color of coagulated blood is washed out, while casein forms the basis of cheese.

Gelatinoids. — These form the nitrogenous substances of bone, skin, and cartilage, and the connective tissue of the animal body. They may be extracted by boiling with water; the resultant product is glue. Their composition is similar to the albuminoids, in that on the average they contain about sixteen per cent nitrogen, the cartilage containing less, and the bones and skin more, nitrogen than the albuminoids.

Horny Matter. — The hair, horn, hoofs, claws, nails, wool, and feathers are constituted mainly of horny matter.

The composition of horny matter is quite uniform, and is similar to albumen in content of nitrogen, though containing more sulphur.

Inorganic Matter or Ash. — The ash constitutes a very small part of the total weight of animals. It ranges from one and eight-tenths per cent to three per cent in swine, and from four and five-tenths per cent to five per cent in cattle. The ash constituents are greatest in lean animals, and least in fat animals. Of the different parts of animals, the dried bones contain the largest portions, reaching fifty per cent in the bones of young animals, and as high as eighty-five per cent in mature animals. Bone-ash consists almost entirely of phosphate of lime. Other very important constituents of the ash are potash, soda, and chlorine.

Animal Food. — The animal body, consisting of the two classes of substances, the nitrogenous and non-nitrogenous, demands the same classes from the food. These latter correspond in kind to those described as contained in the animal body, with the addition of carbohydrates; viz., 1. Albuminoids; 2. Fats; 3. Carbohydrates; 4. Mineral Salts.

Albuminoids. — The albuminoids of a feed include vegetable albumen and fibrin, as well as other substances which resemble in composition the albuminoids of the animal body. The term "protein" is frequently used to designate this class of substances when contained in food.

The various albuminoids vary somewhat in their composition. They are distinguished by their high and quite uniform content of nitrogen; and, though differences exist in them, it is believed that they are quite uniform in their value as nutrients. They are not only the most important of the food compounds, but are indispensable, as they are the sole source directly of the albuminoids in the body of the plant-eating animal.

Other nitrogenous compounds are also contained in small quantities in most plants, though they are relatively unimportant as sources of nourishment.

Fat. — The fat or oil contained in plants agrees closely in chemical composition with that contained in animals. Fats contain a much larger proportion of carbon, and less of oxygen, than the carbohydrates. Fat exists in all plants, and in some seeds, as flax and cotton, in such quantities as to make them of considerable commercial importance as sources of oil.

Carbohydrates. — These agree closely in composition. They consist of carbon, hydrogen, and oxygen only, and derive their name from the fact that the hydrogen and oxygen in them always exist in the same proportions as they exist in water. Cellulose, or woody fibre, and starch are the most abundant of the carbohydrates, though the sugars and gums are also well-known members of this group.

Cellulose is the substance which composes the cell walls or woody part of the plant. It is seldom pure, except in young plants. In trees where great strength is needed, the cell walls become thick and hard, and joined with the cellulose is a harder substance called "lignin." In ordinary farm plants the cellulose exists in greater proportions in the ripe straw, and in the stems and husks of the various plants, than in the seeds. This fibrous material is usually white in color, and is odorless and tasteless. Manufactured flax and cotton, and unsized paper derived from them, are good examples of nearly pure cellulose. It is not readily soluble, and is capable of only partial digestion. It is identical with starch in chemical composition, and may be converted first into dextrine, then into grape sugar, by suitable treatment with acids or alkalies.

Starch is a very abundant substance; it is found in all plants, and in nearly all parts of them. The cereal grains, and the dry matter of root and tuber crops, are especially rich in this substance; and because of its abundance and ease of digestion it is one of the most important of the non-nitrogenous substances. It is readily converted into dextrine and grape sugar by treatment with acids; in fact, the grape sugar or glucose of commerce is largely derived from the starch of corn.

Sugars are of four kinds, — cane, milk, grape, and fruit sugar: these differ but little in composition; all resemble each other in their properties. Cane sugar is derived from sugar-cane and sugar-beet, and milk sugar from the milk of the cow, while grape and fruit sugars usually occur together in the juices of plants, sweet fruits, and in honey. These are all readily soluble in water, and easily digested; and, although occurring in small quantities in ordinary feeds, they are very important, because formed in large quantities from other carbohydrates in digestion.

The gums exist in small amounts in plants used for animal food, and are relatively unimportant food compounds.

Mineral Salts. — These are contained in ordinary fodders in sufficient quantities to supply the needs of the animal body.

The Chemical Analyses of Animal Foods. — By means of chemical analyses the amount of moisture and of the various groups of food compounds described as contained in a food, are determined; viz., albuminoids, fats, carbohydrates, and mineral matter. This grouping is, however, quite incomplete, though serving an excellent purpose in indicating feeding value, and as a means by which a comparison may be made of the various food products.

Water or Moisture. — Water or moisture is determined by drying at a temperature of boiling water.

Crude Protein. — The nitrogenous substances are found by determining the nitrogen, and multiplying the percentage found by the factor 6.25, on the assumption that all of the nitrogen is in the form of albuminoids, which contain on the average sixteen per cent of nitrogen. The substance found in this manner is called "crude protein." In many cases, however, the nitrogenous substances, not in the form of albuminoids, as amides and amines, and contained in considerable amounts in immature plants, are determined separately and deducted from the total crude protein found, in which cases the results are stated as "true albuminoids" and as "non-albuminoids."

Fat. — The content of fat is determined by extracting with ether, or other solvents, and the result is stated as crude fat or extractive matter, since other substances, as gums and coloring matter, are extracted to some extent along with the fat. These may, however, be separated; and when this is done the result is stated as "pure fat."

Crude Fibre or Cellulose. — Crude fibre or cellulose is determined by boiling the substance with weak acids and alkalies, which dissolve the other constituents contained in it. The residue is called "crude fibre," rather than "cellulose," since it contains all the woody substance of the plant, including "lignin."

Ash or Mineral Matter. — The ash or mineral matter is determined by carefully burning the substance, by which means the organic vegetable matter is completely removed.

Nitrogen-free Extract. — The nitrogen-free extract, which includes starch, sugar, and gums, is usually determined by difference; that is, by subtracting the sum of the water, crude protein, crude fat, crude fibre, and crude ash from one hundred. This group, together with the crude fibre, makes the total carbohydrates.

Statement of Analysis. — An example of the usual method of reporting an analysis of a food is here given: —

Water	15.3 per cent.
Crude Fat (extractive matter)	3.3 "
Crude Protein (nitrogenous substances)	12.3 "
Crude Ash (mineral matter)	6.2 "
Crude Fibre (cellulose and lignin)	24.8 "
Nitrogen-free Extract (starch, sugars, gums, etc.)	38.1 "

Functions of the Animal Body. — The object of feeding is to furnish material for maintaining life, and for building up the animal body; and each of the compounds of the food exerts a specific function in the process, though alone they are incapable of completely nourishing the body and maintaining life. The change of the food compounds or constituents into similar animal products is accomplished in the living animal by a series of what are termed "nutritive processes or functions," and include digestion, circulation, respiration, secretion, and absorption.

Digestion. — By digestion the solid matter of the food is brought into a form capable of being absorbed by the blood. Soluble food compounds, as sugar, are absorbed without digestion. Digestion is accomplished by means of the digestive organs, — the mouth, stomach, and intestines. In the mouth the food is ground fine and mixed with the saliva, which, besides softening the food, makes it alkaline, and starts a fermentation, which changes the starch into sugar; when the masticated food passes into the stomach — the fourth stomach in the case of animals that chew their cuds — it comes in contact with the gastric juice. This is an acid substance which acts chiefly upon the albuminoids, converting them into substances called "peptones," which are capable of passing through the lining membranes of the stomach. The food remains in the stomach a sufficient time to bring every portion in contact with the gastric juice; it then passes into the first intestine, where it meets with other alkaline secretions, pancreatic juice and bile, which complete the digestion of the starch, albuminoids, and fat. The absorption of the dissolved constituents of the food now remaining takes place finally in the small intestines; the soluble product passes into the blood, which then nourishes the whole body; the undigested portion is expelled in the form of manure.

Respiration or Breathing.— The air taken into the lungs consists chiefly of oxygen and nitrogen. When it passes out it has lost about one-quarter of its oxygen, and also contains a large amount of carbonic acid; the oxygen has been absorbed by the blood, and the carbonic acid carried out is the product of the oxidation or burning of old tissue. By this oxidation the heat of the body is maintained, and kept in a healthy condition.

Excretion.— The products of oxidation of the animal tissues are carbonic acid, water, urea, and mineral salts. The carbonic acid, as already seen, is chiefly removed by the lungs, and to some extent by the skin; the urea and salts are removed by the kidneys, and the water by all of the organs of excretion.

What is Food?— It is thus seen that, in the processes of life the substances contained in the food are changed into animal product, and that this change is always accompanied by loss. Any material capable of replacing this loss, in whole or in part, is called a "fodder" or a "feed;" and any single compound, like albumen or fat, is called a "nutrient." The first essential in a feed is, then, nutritious compounds; these must, however, be palatable, that is, capable of being eaten, and must possess a certain bulk, in order to properly distend the stomach and supply the needs of the animal in this respect, and must be capable of, at least, partial digestion.

Fodders and Feeds.— Common usage has divided animal foods into two classes, coarse fodders and concentrated fodders, or fodders and feeds. By fodders are commonly understood those products whose chief characteristic is bulk; hay, cornstalks, and straw, belong to this class. By feeds are understood the more highly concentrated materials; the cereal grains, buckwheat, peas, and the mill products belong to this class. Thus, fodders may be classified as hay, straws, green fodders, roots, and tubers; and feeds as grains, mill-feeds, and refuse products.

Hay. — Hay produced from the true grasses, as timothy, orchard grass, herd's-grass, and others, is bulky, and is characterized by a high content of carbohydrates, including crude fibre, and a low content of crude fat and crude protein; that made from the clovers, — red, alsike, and crimson, — while also bulky, is much richer in protein than the others. Hay is variable in its composition; its quality depending upon the kind or variety, the character of soil, the stage of growth at time of cutting, and the method of curing. The chief influence of the kind of soil is upon the content of crude protein; the richer the soil, the richer the hay is likely to be in this substance.

Early or Late Cut Hay. — Early cut hay is also richer in crude protein, and poorer in crude fibre, than late cut: for, although an increase in the weight of dry matter may occur, it is chiefly in the substance cellulose, or crude fibre; besides, a material loss of the more nutritious compounds is likely to take place if it is allowed to ripen thoroughly.

The Best Time for Cutting Hay. — The best time for cutting is when the plants are in blossom; since little, if any, food is absorbed from the soil after that period. Hay that has been subjected to frequent rains, and consequent increased handling, suffers great loss, which falls almost entirely upon the most nutritious parts. This is particularly true of clover hay; the leaves, which are subject to loss in handling, contain the highest content of protein, while the stems consist largely of cellulose or fibre. The loss by repeated wetting, which may reach as high as forty per cent of the dry matter, consists almost entirely of the class carbohydrates. Crude fibre suffers but little loss.

Corn grown for fodder, from which such ears as are formed are not removed, corresponds more nearly with hay in composition than with straw and stalks, though containing much less crude fibre.

Straw and Stalks. — These possess, in a greater degree than hay, the characteristic bulk, since the nutritive matter that first existed in the straw has passed into the seeds, which have been removed; the straw is consequently poorer in protein and carbohydrates, and richer in fibre, than good hay, and shows a proportionately lower rate of digestibility. The food compounds in straw after digestion are, however, quite as valuable, and serve their purpose quite as well as those contained in hay. The variations in composition of straw are due to the same conditions as those which affect hay.

Wheat and rye straw are poorer in protein and fat, and richer in the carbohydrates and fibre, and are coarser and harsher, than oat straw. Corn stalks or stover — the stalks from which the ears have been removed — are richer in protein and fat, and poorer in fibre and carbohydrates, than the straws; they compare fairly well with hay in their composition. All of these products, however, if properly cured, serve an excellent purpose in furnishing both bulk and nutritious compounds, and should be utilized.

Green Fodders. — Green fodders are watery in character, though they contain the same proportion of food compounds as the hay made from them, provided no loss occurs in making the hay. The disadvantage of handling the large amounts of water contained in green fodders is frequently balanced by a decreased loss in dry matter, due to handling in a green state, and by an increased palatability and succulence.

Ensilage. — Ensilage is fodder corn, clover, rye, or other green food, preserved in such a manner — usually in airtight buildings called silos — as to retain in large part all of the qualities of the original product. It is highly regarded, particularly on dairy farms, where succulence in a fodder is a matter of importance. The losses due to fermentation occurring in the silo or pit are considerable, and fall chiefly upon the class carbohydrates; though, according to experiments already conducted, the total loss is less by this method than by curing in the field.

Tubers and Roots. — These all contain large amounts of water, ranging from seventy-five per cent in potatoes to over ninety per cent in turnips. Their value as food depends, perhaps, quite as much upon their succulence, palatability, high rate of digestibility, and good effect upon the animal system, as upon the food constituents, which consist almost entirely of carbohydrates.

Cereal Grains. — The grains or seeds of the cereals are the most important of the feeds. They are especially suited for all kinds of farm animals, and for the various purposes of feeding. They are rich in the three groups of food compounds described, are reasonably constant in composition, and possess a high rate of digestibility.

Corn and oats are the cereal grains most largely used for animal food. Of these corn is richer in carbohydrates, and is used to the greatest extent for fattening, and oats almost exclusively as a food for horses, for which it is peculiarly adapted, both in bulk and in proportion of food constituents.

Mill-Feeds and Refuse Products. — These consist of the ground grains of corn, oats, and rye, either singly or mixed, and the residues of grains or seeds after their use for other purposes has been accomplished. The latter differ from the whole-grain feeds in showing a higher content of nitrogenous matter.

Bran and Middlings. — Bran and middlings derived in the manufacture of flour consist of the outer coverings of the grains used, mixed with more or less of the germ, and are richer in fat, protein, and ash than the original grain, the flour containing a much larger proportion of the carbohydrates. They are very useful when fed in connection with the fodders, since they supply in a concentrated form the nutrients usually deficient in these products. The middlings are, on the whole, better than bran, since they contain less crude fibre and more carbohydrates. Both bran and middlings vary somewhat in composition, due to differences in methods of manufacture, and also to variations in the composition of the original product; these variations are, however, less marked than those which occur in the various fodders.

Brewers' Grains. — Brewers' grains, which represent the bran of barley, and malt sprouts, which are the dried germs of the same grain, are derived in the manufacture of beer from barley. The grains when they have served the purpose of the brewer are very wet, containing, on the average, seventy-five per cent of water; in this condition they are somewhat richer in the food compounds than green fodders, and are an excellent feed. They are, however, liable to ferment rapidly, especially in warm weather, which causes serious loss, besides making them unfit for food. The grains, when dried before fermentation sets in, make a wholesome and highly concentrated food, richer in both fat and protein than bran or middlings, and because of their high food value and bulk are an excellent substitute for oats. Malt sprouts are also rich in protein, though poorer in fat than the dried grains.

Gluten Feeds. — Gluten feeds occur as residues in the manufacture of either starch or glucose (grape sugar) from maize or Indian corn, and consist of a series of products, which, when dried, are classed as gluten feed, gluten meal, germ meal, and corn bran. Gluten feed consists of the entire residue; it is quite bulky, and much richer in fat and protein than the original corn. In gluten meal the hull or germ of the corn has been removed, thus largely increasing the content of both fat and protein. It is one of the most concentrated of the nitrogenous feeds, and should be used with great care. The germ meal contains a large proportion of the germ of the corn. It contains more fat and less carbohydrates than corn, and about the same amount of protein. Corn bran usually consists of a mixture of the germ and hulls of the corn. It contains about the same amount of fat and protein as corn, with less carbohydrates and more fibre. It is more bulky than the others. Germ meal and corn bran serve as excellent substitutes for corn meal. Brewers' grains and all of the gluten products contain much less ash than the original barley or corn, the soluble salts being extracted in the process of manufacture.

Hominy Meal. — Hominy meal, a residue in the manufacture of hominy, consists chiefly of the germ and hull of the corn, and corresponds in composition with the germ meal, though much richer in ash.

Cottonseed Meal. — Cottonseed meal is derived in the manufacture of cottonseed oil from the cottonseed. This feed is of two kinds: one in which the hulls of the seed have been removed, in which case it is called "decorticated;" the other called "undecorticated," in which the hulls have not been removed. Both are rich in protein and fat; though the former is far superior as a food, both because of its higher content of nutrients, and its greater palatability and digestibility. The concentration and nitrogenous character of these feeds make them very valuable, particularly when fed in connection with coarse products, though because of their concentration they should be used with great care.

Linseed Meal. — This product, sometimes called oil meal, is a residue from the manufacture of oil from flaxseed, and is of two kinds, called "old process" and "new process;" the former derived when the fat is extracted by pressure, the other when solvents are used. There is but little difference in their feeding value, the larger amount of fat in the "old process" being balanced by an increased amount of protein in the "new process." Both are rich in protein, and show a high rate of digestibility.

Rice Bran, Peanut Meal. — These are also excellent feeds, though not so generally distributed. It should be remembered, however, that all these mentioned, and a number of minor importance, are valuable for their content of fat, protein, and carbohydrates; that many of them represent the more valuable parts of grains; that they are quite as much feeds as if existing in their original natural condition, and should find a place on all well-managed farms.

CHAPTER XIII.

The Digestibility of Fodders and Feeds; Feeding Standards; Nutritive Ratio; The Exchange of Farm Products for Concentrated Feeds.

The nourishment that may be derived from any food depends not only upon its composition in reference to the specific food substances that it contains, but also upon the amount of these that may be digested by the animal.

Digestibility of Fodders and Feeds. — Pure nutrients, as albumen, starch, or fat, are regarded as entirely digestible. These nutrients, however, do not exist in the various feeding materials in a pure state. They are associated with substances that are indigestible, or that hinder their digestibility; hence the entire digestibility of a food is governed both by the purity of the nutrients which it contains, and by the absence of those compounds which prevent the complete action of the digestive fluids.

Seeds of plants, as a rule, contain matter of a readily digestible character; that is, the nutrients contained in them are relatively pure: but they are, in many cases, enclosed in a hard shell, and, particularly if swallowed whole, resist the action of the digestive processes, thus preventing the digestion of the entire seed. In the same manner the nutrients in such products as hay and straw frequently pass through the animal undigested, because they are surrounded by the woody and almost impervious fibre of the cell wall, which prevents the attack of the digestive fluids.

Fodders, therefore, on account of the great proportion of fibre they contain, are less digestible than the finely ground concentrated feeds. The cutting and crushing of the coarse fodders, and the fine grinding of the various grains and seeds, favorably influence the rate of digestibility. The digestibility of a fodder or a feed is also influenced by other conditions, such as the proportion of the three classes of food constituents contained in it; the period of growth at time of harvesting the forage; methods of curing and storage; the kind or breed of animal to which it is fed; the age of the animal, its individual peculiarities; and many other influences of a minor character, all of which should be regarded in the preparation of rations.

The Digestion Co-efficients; Their Derivation and Use. — The relative digestibility of the different products for various purposes under ordinary conditions of feeding, have been determined by actual feeding experiments. Such experiments have been conducted both here and in Europe, and the data derived from a large number are of great service in determining the comparative value of the different feeds. In these experiments, the food and manure are weighed, and the analyses made; and the difference between the total amount of the constituents in the food and in the dung shows how much of each has been digested;

the amount or the per cent digested is called "the digestion co-efficient."

For example, it is found by experiments that clover hay of average quality shows digestible: —

Fat	43 per cent.
Fibre	48 "
Protein	49 "
Nitrogen-free extract	58 "

These figures represent the proportion or pounds per hundred of the various constituents digested, and are used in calculating the digestibility of other samples of the same kind of product of average quality. They are the digestion co-efficients.

The average composition of clover hay is: —

Water	15.3 per cent.
Crude Fat	3.3 "
Crude Fibre	24.8 "
Crude Protein	12.3 "
Crude Ash	6.2 "
Nitrogen-free extract	38.1 "

The calculation of the digestibility of clover hay is, therefore, as follows: —

Digestible Fat	$3.3 \times 0.43 =$	1.42 per cent.
Digestible Fibre	$24.8 \times 0.48 =$	11.90 "
Digestible Protein	$12.3 \times 0.49 =$	6.03 "
Nitrogen-free extract	$38.1 \times 0.58 =$	22.10 "

The digestible fibre has been found to consist of cellulose; hence, in stating the digestibility of a food, the per cent of crude fibre digested is added to the nitrogen-free extract, and the result stated as follows: —

Fat	1.42 per cent.
Protein	6.03 "
Carbohydrates	34.00 "

The Objects of Feeding. — The objects for which we feed are: 1. Simply to maintain life; that is, to replace by food the natural wastes of the body consequent upon the simple exercise of the vital functions, as described in the previous chapter; and 2. To maintain life, and at the same time to increase animal product or work. It is in carrying out the second object that skill and knowledge are required, in order that the use of the food may result in the greatest and most economical production.

Animal Products Differ in Their Character and Composition. — The different results accomplished by feeding, as maintenance of life, the production of milk, flesh, fat, wool, etc., are not only different in their character, but in order to secure them at the least outlay of actual nutrients, different proportions of the digestible compounds contained in feeds must be provided.

In the simple maintenance of life, where there is no gain in flesh, the chief nutrients required are those which best supply the heat and energy necessary to maintain the vital processes; viz., non-nitrogenous substances: hence it is that hay, straw, and stalks, which possess the requisite bulk, and are rich in this class of substances, serve an excellent purpose in the preparation of maintenance rations for cattle and horses.

To secure a product rich in protein, as milk or flesh, the feeds must contain a greater proportion of protein than is necessary when the object of feeding is fat, since the *digestible protein of the food is the sole source of the protein in the body;* while *all the nutrients may contribute to the formation of the fat.* In a young animal, too, a large portion of the nutriment is used in making muscle, tissue, and bone; while in a mature animal the amount of the food constituents that go to form new products is comparatively small, the larger portion being used in maintaining animal heat.

The proportion of the three general classes of food compounds should therefore be different for the various purposes of feeding.

The Proportion of the Food Constituents Required for the Different Purposes of Feeding. — Experiments in feeding animals have been conducted in which all variable conditions have as far as possible been controlled, in order to secure exact data as to the proper proportions in a ration, as well as the amounts of the nitrogenous and non-nitrogenous substances required for the various purposes of feeding. The results of these experiments have led to the fixing of what are termed "feeding standards;" that is, statements of the amounts of digestible protein, or albuminoids, fat and carbohydrates, which appear to be, and are perhaps under average conditions, best adapted to the various conditions of the animal and the numerous purposes of feeding.

Feeding Standards. — The "feeding standards" in most common use are those of the German experimenter Wolff, which are given in the Appendix. For example, his standard ration per day for a milch cow of one thousand pounds live weight, in full flow, requires twenty-five pounds of organic substance, which shall contain 0.40 (*i.e.*, four-tenths) pounds of digestible fat, 2.50 pounds of digestible protein, and 12.50 pounds of digestible carbohydrates. A ration for dairy cows showing these amounts and proportions, or, in fact, any ration conforming to the required "standard," is also called a "balanced ration," and one showing other proportions is called "unbalanced." These expressions are used to indicate that the proportions in the "balanced ration" are such as to insure the best use by the animal of all the food constituents contained in it; while in the "unbalanced ration" the proportions are such as to indicate a waste of one or the other classes of food compounds, because contained in quantities exceeding the needs of the animal. For instance, if a milch cow is fed a ration higher in carbohydrates, and lower in protein, than is indicated by the standard, she will, of necessity, in order to secure the requisite protein, consume more carbohydrates than the system requires, thus entailing a waste of this substance.

The Usefulness of Feeding Standards.—It is evident that the amounts and proportions of digestible food compounds given by the standard are not the best for every cow under all conditions of full milk flow; for different cows differ not only in their capacity to utilize food, but also in the amount of milk produced when in full flow. The same holds true of standards for other animals; that which is the best for one may not be the best for another.

Feeding standards are, therefore, mainly useful as guides in the selection of food products for the preparation of rations; and, though they should not be regarded as positive rules, the experiments conducted in connection with the experience of practical feeders indicate that the amount and proportion of the digestible food

160   FIRST PRINCIPLES OF AGRICULTURE.

compounds given by the standard can be followed with great advantage. In many cases, too, while the proportions given by the standards would furnish the greatest return for the amount fed, other proportions, because of the prices of feed, may give the largest money returns to the feeder. The use of feeding standards must be accompanied with judgment on the part of the feeder regarding the individuality of the animal, and the character of feeds and their cost, as well as the object of feeding. That is, animals must be fed as individuals, with peculiarities of appetite, digestion, and assimilation, not as fixed machines.

Nutritive Ratio. — In a ration for simple maintenance, the proportion of the fats and carbohydrates together may be greatly in excess of the digestible protein; while for the production of milk, or of flesh products which are rich in albumen and casein, the direct and only source of these compounds, viz., digestible protein, should be proportionately increased. The proportion of the one class of substances to the other is called "nutritive ratio," and is obtained as follows: —

The sum of the digestible carbohydrates and two and one-fourth times the digestible fat is divided by the digestible protein in the ration; the quotient gives the nutritive ratio. The calculation of the nutritive ratio of clover hay, from the analysis and digestibility given on page 156, will serve as an example of the method: —

```
        Digestible Fat, × 2¼  . . . . . . . . . . . =  3.19
        Digestible Carbohydrates  . . . . . . . . . = 34.00
                                                     ─────
                                                     37.19
        Digestible Protein  . . . . . . . . . . .    6.03
```

The nutritive ratio is 6.03 to 37.19, or 1 to 6.2; that is, one part of digestible nitrogenous substances, sometimes called "flesh formers," to 6.2 parts of digestible non-nitrogenous matter, or " fat formers."

Wide vs. Narrow Rations.—If the quantities of digestible fat and carbohydrates are large relatively to the protein, this number will be large, and the ration is called a "wide ration." If the quantities of digestible fat and carbohydrates are relatively small, the quotient is a small number, and the ration is a "narrow" one. A ration where the nutritive ratio is much more than 1 to 6 may be called a "wide ration," if much less it may be called a "narrow ration."

Very few natural feeds conform closely to the standards given for the various purposes of feeding. Coarse farm products show a very wide nutritive ratio, and are only well adapted for maintenance; while many of the concentrated feeds show a much narrower nutritive ratio than is called for by the standards, even for the production of milk or flesh. This makes it necessary, in order to economically use food products, to combine those rich in carbohydrates, or possessing a wide nutritive ratio, with those rich in protein, and possessing a narrow ratio.

The Preparation of Rations.—The preparation of rations requires, however, more than a simple combination of nitrogenous and non-nitrogenous foods in such a manner as to secure the proper nutritive ratio. The bulk of the ration, as well as the palatability and digestibility must be regarded; there must be sufficient bulk to properly distend the stomach; food too concentrated in character, though supplying the requisite nutrients, causes an uneasy and unsatisfied feeling, and it is evident that bulk should be different for the cow than for the horse or pig. For milch cows it may consist, in large measure, of straw, which is partially digestible, and usually palatable; while for the horse or pig an abundance of straw would serve a less useful purpose, because of the smaller size and different formation of the stomach, and because in the latter the food is not re-masticated, as is the case with the cow.

Palatability is also an important feature, since the amount of animal product secured is largely dependent upon the amount of food the animal can be made to consume over and above that necessary to maintain life. Too much indigestible matter must also be avoided; since it is liable to disarrange the system, because of the extra work required to properly dispose of it.

To insure the minimum waste of nutritious matter, such coarse products as corn-stalks and straw, which in their original state are not readily and completely eaten by animals, must be cut, the coarser and finer portions intimately mixed, and feeds of known relish added.

The following examples show how various fodders and feeds may be combined in order to secure such proportions of the digestible food compounds as experiments and experience have taught us are well adapted to the purposes indicated, while at the same time possessing the features discussed in reference to bulk and palatability. The tables in the Appendix furnish the data necessary for the calculation of digestibility and nutritive ratio: —

FODDERS AND FEEDS.

Daily Rations Based Upon 1,000 lbs. Live Weight.

FOR GROWING CATTLE, 6–12 MONTHS' OLD.

1.
15 pounds Corn Stalks.
10 " Wheat Bran.
3 " Linseed Meal.

2.
20 pounds Clover Hay.
6 " Corn Meal.
2 " Linseed Meal.

FOR FATTENING STOCK.

1.
10 pounds Timothy Hay.
8 " Oat Straw.
5 " Corn Meal.
6 " Wheat Bran.
3 " Cottonseed Meal.

2.
10 pounds Corn Stalks.
6 " Wheat Straw.
6 " Wheat Middlings.
3 " Linseed Meal.
5 " Corn Meal.

FOR MILCH COWS.

1.
5 pounds Timothy Hay.
5 " Shredded Corn Stalks.
6 " Corn and Oat Meal.
6 " Wheat Bran.
2 " Cottonseed Meal.
8 " Roots.

2.
10 pounds Mixed Hay.
4 " Corn Meal.
4 " Malt Sprouts.
3 " Wheat Bran.
2 " Linseed Meal.
1 " Cottonseed Meal.

HORSES FOR FARM WORK.

12 pounds Timothy Hay.
6 " Corn.
4 " Oats.
1 " Linseed Meal.

HORSES FOR ROAD WORK.

10 pounds Timothy Hay.
10 " Oats.
6 " Wheat Bran.
2 " Linseed Meal.

These rations, however, are intended to show chiefly good proportions of the various materials; the amounts should be adjusted by the feeder to best meet the requirements of the individual animal. Care in respect to feeding is quite as important as the selection of the various products. The amount and kind of concentrated feed are also matters of importance. The highly nitrogenous cottonseed, linseed, and gluten meals must be fed in small amounts, ranging from two to four pounds per day for milch cows; while bran, dried brewers' grains, and corn meal, can be safely fed in larger quantities. It is desirable, when possible, to make the ration consist of a small quantity each of a number of feeds, rather than a larger quantity of one or two; since it usually adds to the palatability, lessens the danger of overfeeding, and permits a frequent change of diet.

Practical Methods of Using Balanced Rations. — Where the appliances for making weights at each feed are not at hand, and it is preferable to measure, the different materials should be weighed at least once, and the relation between a certain weight and a certain bulk ascertained. The weight or measure of feed for a day's ration for a herd may be mixed together in the proportions given, and in feeding they should be distributed in such a way as to give animals of different weights and capacities for using food that amount best adapted for them. In feeding dairy animals, where there are a number of dry cows, the mixtures for each lot would better be made separately. For horses, the grain or feed rations for work and maintenance may each be mixed in considerable quantities, and placed in separate bins.

Exchange of Farm Products For Concentrated Feeds. — A comparison of the nutritive ratio of the various natural farm products, — hay, grain, straw, and corn stalks or fodder — and of the various feeding standards, as already indicated, shows that, with few exceptions aside from the maintenance ration, these contain a decided excess of carbohydrates, or, in other words, the nutritive ratio of the various farm products is too wide when the purpose of feeding is increase in animal product. This is even more striking if only the coarse products — hay, straw, stalks, and corn — are retained by the farmer for his purposes; hence, in the purchase of feeds that shall balance rations made from home-grown produce, those should be selected which show a narrow nutritive ration or an excess of protein.

Actual Practice is Often Wasteful. — In too many cases in actual practice, in order to enrich the ration fed, constituents are added that are already in excess; and hence, while an increase in result may be obtained, there is also an increased waste of valuable constituents. To add corn meal, an excellent food product, to corn fodder or corn stalks, — that is, carbohydrates to carbohydrates, — in the preparation of rations for dairy cows, may increase the product of the dairy, but it is by virtue of an increased total consumption of food, because of the more concentrated character of the ration, rather than by an economical use of the constituents. A waste of food is only warranted when it is cheaper to waste than to utilize.

Economy in Selling Grain and Buying Feed. — In many cases, too, the cost of the nutrients in the commercial concentrated feeds is much less than is secured by the farmer for the same nutrients in whole grains; under such circumstances, to sell the larger part of his grain and hay crops, and to purchase in return those feeds which will enable him to utilize his coarse products, like straw and corn stalks, to the best advantage, is a desirable practice. It must be remembered that the waste or refuse feed products usually consist of parts of grain, and hence, so far as nutrients go, are quite as serviceable as the original products.

Fertility in Feeds. — A fodder or feed has a fertilizing value by virtue of the manurial constituents — nitrogen, phosphoric acid, and potash — contained in it. Corn, oats, hay, wheat, or other crops, when sold from the farm, carry with them a certain portion of these constituents; and the sale of these products continued for a long time must result in the exhaustion of the soil. If they are returned to the land in whole or in part, they will aid in the growth of other plants, and the time of exhaustion is postponed.

The Relative Fertility of Fodders and Feeds. — Fodders are much less valuable as direct manures than the feeds, — first, because they contain less of the essential constituents; and second, because, from their woody character, their decay is less rapid, though both are valuable as indirect manures, because of their high content of organic vegetable matter. The direct fertilizing value of a feed is largely measured by its content of nitrogen; though the ash constituents, phosphoric acid and potash, are also of considerable importance.

Mill-Feeds vs. Commercial Fertilizers. — Fine-ground mill-feeds, though less concentrated, are quite as good sources of available organic nitrogen as the best commercial forms furnishing that element; the phosphoric acid is less valuable than when contained in forms completely soluble in water, because decay must take place before it becomes available to the plant; while the potash, which is largely soluble in water, is regarded as equivalent in value to that contained in forms free from muriates. Of all the feeds we have, cottonseed meal is the richest in fertilizing constituents. In the Southern States it is used directly as a manure, and even in the Northern States it is one of the cheapest sources of organic nitrogen. Linseed meal, malt sprouts, dried brewers' grains, and a number of others, are much richer in nitrogen than the average commercial fertilizer, though their higher relative cost forbids their direct use as nitrogenous manures.

Refuse Feeds are Rich in Fertility.— Feeds, therefore, in addition to their food value, possess an actual and sometimes a considerable value as a fertilizer. This point should be carefully regarded in their purchase, and especially as bearing upon the point already emphasized : viz., the exchange of home-grown produce for them ; for it is an important fact that the crops grown which contain the highest amount of carbohydrates, and thus in many cases the most desirable to dispose of from the standpoint of economical feeding, are those which contain the smallest amounts of the fertilizing constituents; while those commercial feeds containing protein in large amounts — the substance usually deficient in home-grown produce — are rich in nitrogen, and often richer also in ash constituents. Hence it is that on farms where stock is kept, a judicious exchange of farm products for concentrated feeds may result in soil improvement, rather than soil exhaustion, even with the continuous sale of crops. It is a fact, too that the prices of feeds, as well as the crops sold, are governed entirely by market conditions; that is, no account is made of the fertility value, hence the fertilizer constituents gained by this exchange are a clear gain to the farmer.

A good example of the advantages of a careful observation of these points is shown by a comparison of corn meal and wheat middlings, valuable food products, which contain practically equivalent amounts of total digestible food, and which usually cost about the same price per ton: —

	Nitrogen.	POUNDS OF Phosphoric Acid.	Potash.
One ton of Corn Meal contains	33	14	8
One ton of Wheat Middlings contains	50	28	14
Excess in Middlings	17	14	6

That is, by the exchange of one ton of corn meal for one ton of wheat middlings, there is not only no loss of fertility, but a gain in mineral constituents nearly equivalent to that contained in one ton of the corn meal. The use of tables in the Appendix will enable the student to make correct comparisons of the fertility values of the chief farm crops and purchased feeds.

Manurial Value. — The direct fertilizing value of a ton of feed is, however, greater than the manurial value of the same; since, in feeding, a portion of the fertilizing constituents is retained in the animal itself or obtained in animal product, the amount depending upon the kind of animal, and the object of feeding.

The quantity of nitrogen and ash constituents voided in the manure of a grown animal, neither gaining nor losing in weight, will be nearly the same as that contained in the food consumed. In case animals are increasing in size, producing young, or furnishing milk or wool, the nitrogen and ash constituents in the manure will be less than in the food, in direct proportion to the quantity of these substances which have been converted into animal products.

The data secured in experiments by Lawes & Gilbert, at Rothamsted, England, contained in the following table, are perhaps fairly representative of the amounts of fertilizer constituents retained from the feeds in the various animal products: —

	PERCENTAGE OF NITROGEN.				PERCENTAGE OF ASH CONSTITUENTS.	
	Obtained as Animal Product.	Voided as Solid Excrement.	Voided as Liquid Excrement.	In Total Excrement.	Obtained as Live Weight or Milk.	Voided as Excrement or Perspiration.
Horse at Rest	None	43.0	57.0	100.0	None	100.0
Horse at Work	None	29.4	70.6	100.0	None	100.0
Fattening Oxen	3.9	22.6	73.5	96.1	2.3	97.7
Fattening Sheep	4.3	16.7	79.0	95.7	3.8	96.2
Fattening Pigs	14.7	22.0	63.3	85.2	4.0	96.0
Milking Cows	24.5	18.1	57.4	75.5	10.3	89.7

Manurial Constituents should be saved. — The amount of fertilizer constituents voided in the manure may, however, be greater than the amount actually present when the manure is used, because of the losses that may occur between the time it is produced and the time it is applied to the land; hence the precautions given in a previous chapter concerning the care and management of manures should be carefully followed.

CHAPTER XIV.

Principles of Breeding; The Pure Breeds of Farm Stock.

The various breeds of horses, cattle, sheep, and swine, are the result in large measure of attempts to secure the best. A single breed of cattle, for instance, could not meet in the best manner all the requirements that the numerous conditions now demand. We must have animals adapted for a definite specific production, rather than for general purposes; viz., work, speed, endurance, butter, milk, meat, wool, and fat. A distinct breed is one which possesses distinct characteristics of color, form, and habit, which are transmitted without material change to the offspring. The best breed is that which best meets the demands in any specific case.

Principles of Breeding. — Breeding is an art rather than an exact science, though it is scientific in that it is based upon scientific principles or natural laws. These must be observed in order both to attain and to retain the specific characteristics desired.

Heredity, or the law that "like begets like," is the most important. It is regarded as the corner-stone of the art. This law applies not only to outward form, but also to the entire characteristics; animals inherit the qualities, habits, and tendencies of their parents, both good and bad. The results of this law of descent are observable on all sides, both in the lower animals and in man. Certain families have certain peculiarities or habits, good or bad, that are directly traceable to their ancestors; that is, animals bred true to a given idea or type for a long time acquire fixed characters or peculiarities, which they are capable of transmitting unimpaired to their offspring.

The law of heredity is, however, only true in a general sense. It is not absolute; if it were, the improvement of live stock would be impossible. Certain other natural tendencies are constantly active to modify the law of heredity.

Atavism, or Reversion, is the name given to qualities or habits in the offspring which were not possessed by the immediate parents, but which were possessed by some remote ancestor. This law of atavism, or tendency to go back to an original type, is familiar to all breeders, and frequently causes annoyance, particularly where fine points in breeding are regarded as important. Many instances are recorded in the various herd-books of the appearance of calves with a color totally different from that of their immediate parents, and the appearance of an animal with horns is a quite common occurrence among the recognized hornless breeds of cattle.

Variation is a tendency in the offspring to be unlike their parents. It is by virtue of this law of variation, which is readily influenced by artificial conditions, that rapid changes in types may be effected and new breeds formed.

The great gain in maturity and meat-producing qualities of certain breeds of cattle is due in a great measure to better methods of feeding, in connection with greater care in breeding. This influence has been so marked as to give rise to the common expression that "feed makes breed." Habit also exercises a marked influence in the development of valuable characteristics, a striking example of which is observed in the breeds of milch cows. The habit of giving milk, which has been encouraged for a long time, has caused a change in the structure and function of the animal. In the early breeds the tendency to secrete milk was not a prominent characteristic; it was only sufficiently developed to satisfy the demands of the young.

Prepotency, or the superior influence of one parent over the other in determining the character of the offspring, is also a principle which exerts a decided influence in the development and improvement of distinct breeds. Certain breeds, as well as certain animals, possess this characteristic in a marked degree; that is, the offspring closely resemble this particular breed or animal, whatever may have been the character of the other parent. Among American trotting-horses, Rysdyk's Hambletonian showed this power of individual prepotency in a remarkable degree; his get, as a rule, not only resembling him in color, form, gait, temper, vigor, and endurance, but in nearly every conceivable quality that he possessed.

Lineage. — Various terms are used to express the birth and descent of animals. "Pure-bred," "thorough-bred," and sometimes "full-blood," are terms used to indicate animals of a distinct and well-defined breed. "Pure-bred" is the only strictly correct term; "thorough-bred" is the name of a distinct breed of English race-horses, while "full-blood" hardly expresses the idea.

Cross-Bred refers to animals produced by breeding together distinct breeds; for example, the offspring from the breeding together of pure-bred Shorthorn and Jersey cattle, or of pure-bred Berkshire and Chester white swine, are cross-breeds.

Grades are the product of a cross between a pure-bred and a native. The offspring of a pure-bred Jersey sire and a cow of no fixed type is a "grade" Jersey; while a "high-grade" animal is one in which the blood of a pure breed is in excess. The offspring of a "pure-bred" Jersey and a "grade" Jersey is a "high-grade."

In-and-in Breeding means mating animals that are closely related to one another. This method, as practised by various breeders, differs both in respect to the degree of relationship, and to the continuation of the practice. Authorities have defined the term as applying "only to animals of precisely the same blood, as own brother and sister," and also as "pairing of relations within the degree of second cousins twice or more in succession."

In-and-in breeding, carefully followed, permits the rapid establishment of a uniform breed; if carried too far it is likely to be accompanied by a loss in size and constitutional vigor, though the method is followed to a greater or less extent by all breeders.

Pedigree is the record or statement of the ancestors of an animal, and is usually registered only in the case of pure breeds; it is useful as a guide in tracing inherited qualities. It is the custom when a distinct breed has been established to issue a herd or flock book. The methods of recording, and the rules governing the registry of pedigrees, are adopted by the various Breeders' Associations; and while they differ somewhat in regard to form, the object is the same: viz., to put in permanent form a true record of the lineage, character, and performance of the individual.

Value of Pure Breeds. — Pure breeds, as already stated, are the result of attempts to secure the best animal for some specific purpose; and they can be relied on not only to accomplish that purpose better than any other, but also to produce young that possess the same qualities. For example, a specific milk or beef breed of cattle, or wool or mutton breed of sheep, will produce either milk or beef, wool or mutton, better than milk *and* beef, and wool *and* mutton. Whereas, in the common or native stock — a mixture of many breeds — a uniformity in production cannot be depended upon, as the type and character are not fixed, though individuals may possess superior qualities.

The pure breeds, too, because of their fixity of character and prepotent power, are extremely valuable in improving native stock. The offspring of a pure-bred sire and a native dam will possess in greater degree the character of the pure-bred sire than that of the mixed-bred dam. A pure-bred sire is, in such cases, more than one-half of the herd or flock.

Breeding as a Business. — To be a successful breeder of live stock requires large capital, broad knowledge and experience, great patience, and a close attention to the details of the work. As a rule, breeding is more successful when conducted as a distinct branch of farming than when added to the work of the general farmer. A farmer may, however, greatly improve his stock by careful attention to the principles which govern in breeding, and, so far as the products of his herds and flocks are concerned, get much better results than are possible from common stock. The best animals for general farmers, for instance, are without doubt "high grades," produced from crossing pure-bred sires with good common stock. For this work a knowledge of the conditions of the farm, coupled with the knowledge of the characteristics of the leading breeds, should serve as a guide for the selection of that breed which shall best fulfil the conditions. In this work it is quite as necessary to have a sire of a high degree of excellence as in herds of pure-breds; for the stronger the fixed qualities of the sire, the more likely is he to impress them upon his offspring, and to overcome the tendencies inherent in the common dam.

Cross-Breeding. — The crossing of the pure breeds has not always proved successful, since the inherited tendencies are too strong to admit of an equal mingling of the characteristics of the parents; however, many such crosses have proved satisfactory. In this work, breeds should be selected which possess certain qualities in common, rather than those possessing distinctly opposite characters; for instance, the breeding together of a very small and a very large animal is not so likely to result in offspring possessing the best characteristics of both, as if the parents were more nearly alike in this and other respects.

Breeds of Horses. — The distinct breeds of horses are classified as draft breeds, heavy carriage breeds, thorough-breds, American saddle-horses, American trotting-horses, and pony breeds.

Draft Breeds. — Draft breeds, or large, heavy horses, —

of weight ranging from one thousand six hundred to two thousand pounds, and specially adapted for heavy teaming, — include the Percheron, and French and Norman Draft, originating in France; the Clydesdale, native of Scotland; the English Shire and Suffolk Punch, of England; and the Belgian Draft, native of Belgium. Of these, the Clydesdale is perhaps more suitable than the others for heavy farm work. They are fast walkers, intelligent, gentle, and easily broken.

Heavy Carriage Breeds. — The Cleveland Bay, French and German Coach and Hackney, constitute this class. These breeds are large, active, and stylish, and, while bred for heavy coaching, are well adapted for road or farm work. The "grades" or "high grades" of the Hackney, particularly, are highly valued, both for the farm and for general driving. The use of sires of all these breeds has proved of great value in improving our common stock of horses.

Thorough-Breds and American Saddle-Horses. — "Thorough-bred" applies only to the English running-horse — the American "thorough-breds" being either imported from England, or the descendants of horses so imported. This breed had its origin in the East, and is the improved breed of the Arab, Turk, or Barbarian. It is the oldest as well as the most noted of all breeds. The best qualities of many breeds are due to a greater or less admixture of thorough-bred blood.

The American saddle-horse is a newly formed breed, which was originated by a judicious mingling of the blood of the thorough-bred with the pacer.

The American Trotting-Horse. — This class, which

traces directly to the thorough-bred, is not yet recognized as a distinct breed, though it is better known here than are those of the pure breeds.

Pony Breeds. — These consist of the Shetland, Welsh, Exmoor, Mexican, and Indian, each possessing special characters, form, and habits.

Breeds of Cattle. — These are usually classified as dairy and beef breeds, though many are regarded as possessing both dairy and beef qualities in a marked degree.

Dairy Breeds. — These may be further classified as butter and milk breeds. Chief among the butter breeds are the Jersey and Guernsey, natives of the Channel Islands of the same name situated near the north-west coast of France.

Jersey. — This breed is the most noted of all the dairy breeds, both for its general elegance of proportion and appearance and for its excellent qualities. For butter-making it is not excelled. It has been brought to its present perfection by very careful methods of breeding, in which one idea, viz., butter, is constantly followed. It is small, and possesses a rather delicate constitution, and is thus not adapted to rigorous conditions of climate and careless handling.

Guernsey. — The Guernsey is not so general a favorite as the Jersey; it is larger and coarser, though the texture of its skin is extremely delicate. It is distinctly a dairy breed, and is a much deeper milker than the Jersey; the butter product is also richer in color and of better texture. A crossing of the Jersey or Guernsey upon our common stock is extremely useful in improving their butter qualities.

Milk Breeds. — The chief milk breeds are the Ayrshire, Holstein-Friesian, and Shorthorn; though the Dutch Belted, Brown Swiss, Devon, and a few others have attained considerable prominence in certain localities.

Ayrshire. — This breed is traceable to the county of Ayr in Scotland. Their chief characteristic is their excellent milk, good in quality and large in quantity. The prevailing color is brown and white; it is extremely hardy, active, and well adapted for mountain districts.

Holstein-Friesian. — The exact origin of this breed is not well established. It is only known that for an indefinite period, anterior to the records of history, there existed a superior breed of cattle in the Duchy of Holstein in North Holland and Friesland. They have been used by the English for two hundred years to improve their stock. In color they are almost universally black and white. Their strong points are large size, deep milkers, and hardy constitutions.

Shorthorn. — The Shorthorn breed was once spoken of as the Teeswater, or Durham. The date since which the breed has had a distinct existence has been disputed, though it was certainly known to have been established in the early years of the last century. The Shorthorns are strong, deep milkers, possess hardy and vigorous constitutions, and a great power of adaptation to changes of soil, of climate, and of pasturage. In many sections, and especially in America, its breeding has been conducted with the sole view of the production of beef; it has also achieved wonderful results through crossing with other breeds. The ranch cattle of the prairies of the West and in Texas have been largely graded up with this breed. Their color ranges from the blood red to the pure white. Owing to their generally valuable characteristics, they more nearly approach the general purpose animal than any other breed, though the Red Polled and Devon are also included in this classification.

Beef Breeds. — The chief distinctive beef breeds are the Herefords, Galloways, and Aberdeen, or Polled Angus; though, as already stated, certain families of the Shorthorn are bred exclusively for beef.

Hereford. — Hereford cattle originated in Herefordshire and adjoining counties in England; they are highly regarded there, and have also met with great favor in the United States. The usual color is a rich, light or dark red, with white face, throat, and chest. The use of the Hereford for crossing with other breeds is not usually attended with as good results as are secured from the use of the Shorthorn.

Galloway. — The Galloway is a polled breed, and derived its name from the province of Galloway in Scotland. The color is universally black, and the hair long and shaggy. This breed has proved extremely valuable for the Western ranges. The animals are easily acclimated, active, and hardy.

Aberdeen Angus. — These are also hornless, and resemble the Galloway in color and form, though they are somewhat less hardy and mature earlier.

Breeds of Sheep. — The Merino, Southdown, Shropshire, Hampshire, Oxfordshire, Cotswold, Leicester, Lincoln, and Horned Dorset are the leading breeds.

Merinos. — These now include many distinct strains.

They are the most widely known of all the breeds of sheep in America; they are bred almost exclusively for their fine wool, for which purpose they are unexcelled by any other breed. Their mutton qualities, while much improved by careful breeding, are not of a superior character. They are hardy, well adapted to warm climates, and the rams have been extensively used for breeding up the flocks in the South-western States.

Southdown.—Next to the Merino, the Southdown is the most extensively distributed breed in the United States. In size they are above the medium, and for the production of mutton take first rank. The ewes are prolific, and the lambs are vigorous and hardy.

The other "down" breeds, viz., Shropshire, Hampshire, and Oxfordshire, resemble somewhat the Southdown in mutton and wool producing qualities, though showing differences in size and in their ability to thrive under varying conditions.

The Cotswold, Leicester, and Lincoln are bred chiefly for their long wool. They are larger, and, as a rule, less prolific than the various down breeds; they are extensively used in crosses to improve size.

Horned Dorset is an old and well-established breed in England, where it originated in the shire of Dorset. It is not largely distributed in America. In size these sheep are above the medium. For the production of early, fat lambs this breed has no superior. With proper management they may be made to breed at all times of the year, are very prolific, dropping a large portion of twins, and are good nurses. This breed should occupy an important place here in the production of early lambs.

Breeds of Swine are usually divided into classes according to size. The large breeds, which are well distributed in America, include the Berkshire, Poland-China, Duroc, or Jersey Red, and Chester White, and the medium and small breeds, the Improved Berkshire, Cheshire, Small Yorkshire, Essex, and Suffolk.

The larger breeds are more generally distributed in the corn-growing States of the Central West, and are well adapted for supplying the large pork-packing houses located there; while the smaller breeds are more generally distributed in the more thickly populated districts, and, because of their early maturity, are better adapted to supplying the demands for light pork for immediate consumption.

The Duroc, or Jersey Red, the Chester White, and the Poland-China are American breeds. The Berkshire, Cheshire, Yorkshire, Essex, and Suffolk are English breeds.

The chief characteristics of a good hog are early maturity, quietness of disposition, and small percentage of loss in dressing.

CHAPTER XV.

The Products of the Dairy; Their Character and Composition; Dairy Management.

The distinct products of the dairy are milk, cream, butter, and cheese; and the waste or by-products, skim-milk, buttermilk, and whey.

The primary purpose of the milk of the cow is to feed and nourish her young. The secretion or formation of a larger quantity than is required for this purpose is, therefore, an acquired character, and is the result, in large measure, of artificial conditions.

Milk is a Food in the fullest sense. It not only contains the nutrients necessary to sustain life and to cause growth, viz., fats, albuminoids, carbohydrates, and mineral salts, but these exist in such a form as to be readily digested. Milk also possesses physical properties which distinguish it from other products. It is a white fluid, throughout which the fat is distributed in the form of small globules. The fat is lighter than the remainder of the fluid, which contains the albuminoids, carbohydrates, and salts in solution; hence, on standing, the fat globules rise to the surface. This property is taken advantage of in the preparation of the products, cream and butter.

Fat of Milk, or Butter-Fat, consists of a number of distinct kinds of fat, the chief of which are palmatin, stearin, olein, and butyrin. These may be classified as fixed, that is, those which remain on heating, and as volatile, those which may be driven off by heat: fixed fats are also of two kinds, solid and liquid.

The volatile fats affect the flavor of dairy products more than the fixed, and it is the proportion of liquid fat (olein) which affects the solidity of butter. The liquid fats increase with succulent foods, and the solid with dry foods.

Albumen and Casein.— These two substances constitute the chief albuminoids of milk; and while they resemble each other in composition, they possess different properties. The albumen, which is contained in small amounts not usually exceeding one-half per cent, is coagulated by heat and not by acids, while casein is coagulated by acids and not by heat. This property of casein is very important in the manufacture of cheese.

Milk Sugar, called by chemists "lactose," possesses practically the same food value as other sugars. It differs from cane sugar in appearance and in its properties. When it is crystallized, it is very hard, and it does not possess as high a sweetening power.

Ash, or Mineral Salts, consists of phosphates of lime, magnesia, and iron, and chlorides and sulphates of soda and potash.

Average Composition of Milk.— Milk is not a product of fixed composition. Both the total amount and the proportion of the constituents are influenced by a variety of conditions, the chief of which are: breed of the animal; her age, health, and individuality; the method of feeding and kind of food; period of lactation, and time and season of milking. Of the constituents, fat varies more than the others, though each may vary sufficiently to cause serious differences in the composition of the products made from milk. The accompanying analysis fairly represents the average amounts and proportions of the constituents in normal milk: —

Water	87.50 per cent.
Fat	3.50 "
Casein and Albuminoids	3.75 "
Milk Sugar	4.50 "
Ash	0.75 "
	100.00

This average composition of milk has served as the basis in many States for the enactment of laws to prevent watering and other forms of adulteration. It must be remembered, however, that normal or whole milk will show wide variations from this standard in both directions; that is, it may be very much richer or very much poorer. The solid matter in milk is called "Total Solids."

The Influence of Breed. — It has already been stated that cattle are divided into two classes: on the one hand, those in which the tendency to secrete milk is largely developed; and on the other, those in which the tendency to form flesh and fat has been especially encouraged. The result of this careful selection is the formation of a distinct milk type, in which the width and depth of the hind part of the animal and the udder are especially prominent features.

The dairy breeds are, however, further classified into milk and butter breeds; that is, those which give a large quantity of average quality, and those which give a smaller quantity of a higher quality. The following table of averages, the result of experiments conducted at the New Jersey Experiment Station, with representatives of the leading dairy breeds, shows their relative yields, and the composition of the milk: —

Average Yield and Composition of Milk of Different Breeds.

HERD.	Yield per day in quarts.	Percentage of					
		Water.	Total Solids.	Fat.	Casein.	Sugar.	Ash.
Ayrshire	9.0	87.30	12.70	3.68	3.48	4.84	0.69
Holstein-Friesian	11.0	87.88	12.12	3.51	3.28	4.69	0.64
Shorthorn	9.0	87.55	12.45	3.65	3.27	4.80	0.73
Average	9.7	87.58	12.42	3.61	3.34	4.78	0.69
Guernsey	8.7	85.52	14.48	5.02	3.92	4.80	0.75
Jersey	8.4	85.66	14.34	4.78	3.96	4.85	0.75
Average	8.6	85.59	14.41	4.90	3.94	4.83	0.75

While these results are not absolute, it is evident that there is a distinct classification of breeds based upon the relative yield and quality of milk. The milk from animals which naturally produce large quantities shows average quality, and that from animals which produce a smaller quantity shows a quality considerably above the average. That the content of fat in milk varies more than the other constituents is also distinctly shown in this work. The variations in the composition of milk, due to breed, is, therefore, important in indicating the animals best adapted for the production of a specific dairy product.

The Age and Health of the Animal also affect the composition of milk. As a general rule, the milk of young animals is richer than that of old; this is not positive, however, since much depends upon the vitality, vigor, health, and management of the animals.

The Period of Lactation is the time which elapses between the birth of the calf and dryness, and varies with different animals even of the same breed. During this period the yield and composition of the milk vary. The milk flow is greatest and the quality poorest in the beginning; as the period increases the flow gradually falls off, and, as a rule, the quality improves, though the rate of improvement is dependent somewhat upon food and management. The fat globules are larger at the beginning and smaller at the end of the period.

Colostrum. — A few days elapse after the birth of the calf before the milk is fit for use. The product obtained is called "colostrum," and is especially suited to the needs of the young offspring. It differs from milk in containing a much larger amount of solid matter and ash, and in showing but little sugar.

Milk Drawn at Different Times also differs in composition, though the influence of time of milking is not the same for all animals. In some cases the morning's milk will be greater in quantity and poorer in quality than the evening's milk, while in others the reverse is the case; hence the variation in the milk of a herd is not so noticeable as that from individual cows.

It is also a matter of common observation that the milk first drawn is poorer, particularly in fat, than the "strippings," or that last drawn; frequently the "strippings," or last pint drawn, contain six to eight times more fat than the first pint, while the other constituents, albuminoids and sugar, are more evenly distributed; this variation in composition, in connection with the fact that the fat globules are larger in the "strippings" than in the milk first drawn, indicates that the fat rises in the udder.

The Influence of Food is perhaps greater than any other factor in determining the profit that may be derived from the dairy; its influence is felt, not only on the quantity of milk produced, but on the quality of the products derived from it. A specific breed possesses certain capabilities, the value of which are dependent in large measure upon the food that is supplied.

By proper feeding is meant, not only that the animal should receive a sufficient amount of nutriment in the right proportions, but also that the materials furnishing the nutrients should be clean and wholesome, and free from any substance that may injure the quality of the product.

Pasture and Hay. — Pastures and green foods, for instance, composed only of the true grasses and clovers, are nutritious and wholesome, and can have no injurious effect upon the health of the animal or the quality of the product; while those which include a large number of weeds may not only be dangerous to the health of the animal, but may cause an undesirable flavor in the milk, and an inferior quality of the butter or cheese produced from it. Hay free from weeds, if well made and the desirable properties retained, is an excellent food; but if improperly cured and the characteristic odor destroyed, and so badly stored as to cause it to heat and mould, its feeding will result in a much poorer quality of product.

Coarse Products and Concentrated Feeds. — The coarse products, straw and stalks, and the concentrated feeds composed of the cereal grains and refuse mill products, if clean and sweet, affect the quality of milk only by virtue of the variations in their feeding value, though the character of the products made from it may be influenced to some extent. The feeding of cottonseed, for instance, has a tendency to increase the proportion of solid fat, while gluten meal, on the other hand, is said to increase the proportion of liquid fat.

Useful Succulent Foods, as turnips, swedes, mangel-wurzels, cabbage, etc., also affect the flavor of the milk; and, in order to prevent as far as possible their unfavorable effect, they should be fed immediately after milking. Wet brewers' grains, distillery refuse, and ensilage in an advanced state of fermentation, also exert an unfavorable influence on the quality of the milk.

Changes in Milk. — It is well known that milk from healthy cows, even under good practical conditions of preservation, will remain sweet but a short time; though, if it could be drawn and placed so that no air could come in contact with it, it would always remain sweet. The changes in milk, or tendency to sour, are caused by the entrance into it of ferments, or minute organisms called "bacteria;" and milk possesses in a marked degree those properties which, given a suitable temperature, favor their rapid development.

Good and Bad Ferments. — These bacteria are of two classes, one of which includes those called "friendly," which are necessary or helpful in the making of butter or cheese, and the other, "unfriendly," or those which introduce bad qualities into the milk and its products. The ferments that injuriously affect milk are more abundant in warm weather, in closed buildings, and around decaying matter, than in cold weather and in the open air, because warmth, impure air, and unclean conditions are favorable for their growth, while a low temperature, sunshine, and pure air prevent their rapid development.

Cleanliness is Essential to Good Milk Supply. — Milk from healthy cows, that are fed clean, wholesome food and pure water, and are kept in clean stables, when drawn by clean milkers, and then rapidly cooled and kept in a clean place, will keep longer than that drawn from animals that are poorly fed, improperly housed, and badly cleaned; since in the one case the conditions are such as to reduce the influence of the "unfriendly" bacteria, while in the other the conditions are favorable for their development and introduction into the milk.

It is not only essential that the animal, the stables, the milker, the utensils, the dairy-room or cellar, should be kept as clean as possible, but that all manner of decaying matter about the farm should be prevented or removed. Fermenting foods, bedding consisting of decaying straw or hay, muddy and filthy water in the pastures, are frequently the cause of bad taints in milk. These not only destroy in large measure its good qualities, but render its use dangerous. Furthermore, no milk from diseased animals, or that which has been exposed to the germs of infectious human diseases that may be carried and introduced by milkers, or diseases which can be introduced through the animals themselves by means of contaminated water supply, should ever be offered for sale direct as milk, or indirectly as butter or cheese.

Cream consists of the fat globules mixed with more or less of the other constituents of milk. Its richness in fat depends upon the quality of the milk from which it is derived, and upon the method used in creaming. It is, therefore, not a product of uniform composition; in fact, it is much less uniform than milk, its content of fat — the chief constituent of value in it — ranging from as low as ten per cent to as high as forty per cent.

Its purchase or sale, either as food for families or for the production of butter at creameries, should be based on the actual content of fat rather than its volume or measure.

Systems of Creaming. — These are divided into two classes, — first, the setting systems, in which the cream rises under natural conditions; and second, centrifugal systems, in which mechanical force is used.

The simplest setting system is the open-air shallow pan; it is also the most common, but it does not give the best results. In order to get the largest quantity of cream by this method, the milk has to stand too long, which endangers the quality of the butter; besides, the long exposure to the air induces rapid changes and souring, which render the skim-milk less valuable as food for animals, and make it unfit for human food. The deep-setting systems permit of a better regulation of the temperature, and of a more perfect protection from the air; while the rapidity and completeness of the creaming is not decreased.

Mechanical separation is more economical of space, time, and labor, and a larger percentage of the fat of the milk is obtained by it than by any other method; besides, perfectly fresh cream and skim-milk can be immediately obtained by this system. This system of creaming has taken the place of the others to a great extent in large dairies and creameries, though the cost of the separator and the power required to run it have prevented its rapid adoption in the small home dairy.

The various large machines have now reached a remarkable degree of perfection; and it is only a question of time before those adapted for the small dairy, both in point of cost and power, will be available.

Butter.—Butter consists of the fat globules of milk gathered into a solid form. Like other products of the dairy, it is subject to wide variations in composition and quality. Good butter should contain at least eighty-five per cent of pure butter fat, not more than twelve per cent of moisture, and less than one per cent of casein. The content of ash, or mineral salts, depends upon methods of salting, though it should not exceed one and one-half per cent.

The yield of butter from a given quantity of milk depends chiefly upon the amount of fat in the milk, and the composition of the product secured. If milk is bought for butter-making, it should be paid for on the basis of content of butter fat, rather than by weight or volume. The yield of butter from a given quantity of fat in different lots of milk also varies slightly, since the fat in all milks cannot be uniformly recovered as butter even under uniform methods of treatment. This is believed to be largely due to differences in the size of the fat globules; the larger the globule the greater the proportion of fat recovered; in practice, however, this point is largely disregarded.

The properties of butter which determine its edible quality and appearance are flavor, keeping quality, solidity, texture, and color. These are the result in large measure of the management of the milk and cream, and the method of making the butter. Good flavor, for instance, belongs to some extent to certain of the fats themselves, though largely to the changes which occur in the ripening of the cream, a process of fermentation which can be controlled by the butter-maker. In fact, certain ferments have been discovered and isolated, which, if added to the cream, will give to the butter the delicate flavor so pleasing to the palate. This method of securing uniform quality in respect to flavor is likely to become an important feature of butter-making.

Bad Flavors or Odors may be due to certain foods, as cabbage, poor hay, fermenting brewers' grains, and ensilage, and to such weeds as garlic. Butter will also absorb the odors of foods, decaying substances, etc., with which it comes in immediate contact; hence products possessing distinct flavors should not be stored in the dairy room.

The Keeping Quality of butter is governed to a great degree by the method of making. If the cream is properly ripened and churned, the butter well worked and evenly salted, it will, even under ordinary conditions, retain its original quality for a long time; while if the processes have been carelessly conducted, the buttermilk not completely removed, and unevenly salted, it will soon lose its good qualities; the casein which has been left in it will decay, and cause it to become rancid.

Texture. — When butter is solid, and shows a decided granular structure rather than a greasy appearance when broken, it is said to possess good texture. This is governed by the character of the milk and method of management. Milk which contains large fat globules, as that from the Jersey and Guernsey, will, under the same methods of making, produce butter of a better texture than that from the milk containing small fat globules. Too much handling, and too high a temperature in making the butter, also injure the texture.

The Natural Color of butter is due to a substance in milk called "lacto-chrome." The butter from the distinct butter breeds or their "grades" possesses a better natural color as a rule than that from the milk breeds.

Sweet Cream Butter is used to a limited extent in certain localities; it is made from unripened cream, and is preferred by certain customers because only the original flavors are retained.

Cheese. — Cheese consists of the casein of the milk with more or less of its butter fat; it is of two distinct kinds; viz., whole-milk cheese, which contains all the fat of the milk that can be recovered in the process of manufacture, and skim-milk cheese, where the fat in the milk has been partially removed before it is made into cheese.

Manufacture of Cheese. — The principles involved in the manufacture of cheese are practically the same for the many different varieties. The various operations include the coagulation of the casein, removal of the whey, salting, pressing, and ripening. The composition of whole-milk cheese, as well as its edible qualities, depends upon the composition of the milk used and the methods of manufacture. Recent experiments have shown that, other things being equal, the fat in milk measures the amount and quality of cheese that may be made from it. The best cheese is produced, and the least loss occurs in manufacture, from milk rich in fat.

Good Cheese should possess richness, that is, good proportions of fat and casein, good flavor, keeping qualities, and firmness of texture; and can only be secured by very careful attention to the details in the numerous operations required in its manufacture.

Cheese as Food. — Whole-milk cheese, though varying in composition, is a very nutritious food, since it is rich in the most valuable nutrients, casein and fat. Good products contain as high as thirty-five to forty per cent of fat, twenty-five to thirty per cent of casein, and as low as twenty per cent of water. Skim-milk cheese, though an excellent food, is less valuable; it contains more water and much less fat.

Skim-milk. — This consists of the remainder of the milk after the removal of the fat, and varies in composition according to the completeness of skimming, or separation of fat; as a rule, separator skim-milk is poorer in fat than that derived by other methods. Skim-milk shows less total solid matter and different proportion of the food constituents than whole milk; fat is the most variable constituent. It ranges from two-tenths of one per cent to one per cent; casein and ash are slightly less, while milk sugar is somewhat greater in amount than in whole milk. An average composition would probably show: —

Water	90.00 per cent.
Fat	0.80 "
Casein and Albumen	3.00 "
Sugar	4.90 "
Ash	0.70 "

As a Food, skim-milk, though dilute, is, when sweet, a wholesome and nutritious human food. It is also, both in its sweet and sour state, an excellent animal food, and is especially adapted for pigs and calves; though, because of its highly nitrogenous character and narrow nutritive ratio, it should be used in connection with those of a fatty or carbonaceous nature, which will widen the ratio. Skim-milk and flaxseed meal — which is rich in fat — make an excellent and well proportioned ration for young calves; while skim-milk and wheat middlings, or other products showing a high content of digestible carbohydrates and fat, make a good and economical ration for pigs.

Buttermilk contains the casein and sugar retained in the cream from which butter is made, and such proportions of the fat as are not recovered. It differs but little in composition from skim-milk, and has about the same feeding value, though usually containing more fat and casein, and less sugar; it is also liable to considerable variation, owing to differences in methods of obtaining the cream and of churning.

Whey is the residue from the manufacture of cheese. It is more dilute than the other refuse products, and as a food is chiefly valuable for its content of sugar.

Dairying. — The success and profit of the dairy depend upon a number of conditions, which should be carefully considered. The situation in reference to home supplies, which include water, and natural fertility of soil, access to good wholesome foods, location and character of markets, and the relative profitableness of dairying and other lines of farming, should all be carefully studied before entering upon the business.

The Selection of a Specialty is also important; for, while a series of products may be made, the adoption of a single line usually results in a greater concentration of energy, and hence a better product. This involves a knowledge of the special characteristics of the different breeds, and the principles that govern in their selection, management, care, and improvement.

Testing the Animals. — The profits of the dairy are also governed in large measure by the yield and quality of the milk; hence careful records should be kept of individual animals in these respects. It has already been shown that the yield and quality of the dairy products, cream, butter, and cheese, are measured by the content of butter-fat in the milk; it is, therefore, of the greatest importance that the content of fat in the milk of each animal should be tested. This may be accurately and rapidly accomplished by what are known as semichemical methods; of these the "Babcock Test," devised by Dr. S. M. Babcock of the Wisconsin Experiment Station, furnishes accurate results, and is so simple in operation as to be readily performed by any careful dairyman.

A careful study of the animals in these respects teaches the dairyman the actual value of each, hence only those which are profitable need be kept.

Dairy Products and Soil Fertility. — The relation of the sale of the various dairy products to soil exhaustion is frequently disregarded in the selection of specific lines, though it is a matter of some importance. If whole milk is sold, there is removed from the farm for each ton sold an average of twelve pounds of nitrogen, four and a half pounds of phosphoric acid, and three and a half pounds of potash. If it is manufactured into cheese, the mineral salts are largely retained in the whey, and nitrogen only is removed. If it is manufactured into cream or butter, and the skim-milk and buttermilk — foods of considerable value — are retained, practically no loss in fertility results. The exact value of these relations will differ with different conditions, such as relative prices received for the various products, the cost of actual fertilizing constituents, and the usefulness of the foods, skim-milk and buttermilk; hence it can, of course, be determined only by the individual dairyman.

The Purchase of Foods and Methods of Feeding are also valuable factors. The feeds should not only be well adapted in themselves, but should be so adjusted to others in the ration as to result in the greatest possible product for the least outlay. The care of the animals should also be kindly, regular, and punctual, and all the processes of the dairy carried out in such a manner as to guarantee the highest quality of product.

APPENDIX.

CONTAINING TABLES

SHOWING THE

Composition of Fertilizing Materials, Farm Manures, Fodders, Feeds; the Coefficients of Digestibility of Various Feeding Stuffs; Fuel Value of Food; Feeding Standards for Different Animals and Different Purposes of Feeding; and the Fertilizer Constituents Contained in the Chief Farm Crops and Concentrated Feeds.

COMPOSITION OF FERTILIZING MATERIALS.

Table I. Nitrogenous Materials.

	POUNDS PER HUNDRED.		
	Nitrogen.	Total Phos. Acid.	Potash.
Nitrate of Soda	15¼ to 16		
Sulphate of Ammonia	19 to 20¼		
Dried Blood (high grade)	12 to 14		
Dried Blood (low grade)	10 to 11	3 to 5	
Concentrated Tankage	11 to 12¼	1 to 2	
Tankage (bone)	5 to 6	11 to 14	
Dried Fish'Scrap	7 to 9	6 to 8	
Cottonseed Meal	6½ to 7½	1½ to 2	2 to 3
Castor Pomace	5 to 6	1 to 1½	1 to 1½

Table II. Phosphatic Materials.

	POUNDS PER HUNDRED.			
	Nitrogen.	Phosphoric Acid.		
		Total.	Available.	Insoluble.
S. C. Rock Phosphate	. . .	26 to 28	. . .	26 to 28
S. C. Rock Superphosphate,	. . .	13 to 16	12 to 15	1 to 3
Fla. Land Rock Phosphate	. . .	33 to 35	. . .	33 to 35
Fla. Pebble Phosphate	. . .	26 to 32	. . .	26 to 32
Fla. Superphosphate	. . .	16 to 20	14 to 16	1 to 4
Bone-black	. . .	32 to 36	. . .	32 to 36
Bone-black Superphosphate,	. . .	17 to 18	15 to 17	1 to 2
Ground Bone	2¼ to 4¼	20 to 25	5 to 8	15 to 17
Steamed Bone	1¼ to 2¼	22 to 29	6 to 9	16 to 20
Bone (dissolved)	2 to 3	15 to 17	13 to 15	2 to 3

Table III. Potassic Materials.

	Pounds Per Hundred.				
	Actual Potash.	Total Phos. Acid.	Lime.	Nitrogen.	Chlorine.
Muriate of Potash	50	. . .	. . .	. .	45 to 48
Sulph. of Potash (high grade)	48 to 52	. . .	. . .	. .	½ to 1½
Double Sulph. of Potash and Magnesia	26 to 30	. . .	. . .	. .	1½ to 2½
Kainit	12 to 12½	. . .	. . .	. .	30 to 32
Sylvinit	16 to 20	. . .	. . .	. .	42 to 46
Cottonseed Hull Ashes . .	20 to 30	7 to 9	10		
Wood Ashes (unleached) . .	2 to 8	1 to 2	30 to 35		
Wood Ashes (leached) . . .	1 to 2	1 to 1½	35 to 40		
Tobacco Stems	5 to 8	. . .	3.5	2 to 3	

Table IV. Average Composition of Farm Manures.

Farm Manures.	Pounds Per Hundred.			
	Nitrogen.	Total Phos. Acid.	Potash.	Lime.
Cow Manure (fresh)	0.34	0.16	0.40	0.31
Horse Manure (fresh)	0.58	0.28	0.53	0.21
Sheep Manure (fresh)	0.83	0.23	0.67	0.33
Hog Manure (fresh)	0.45	0.19	0.60	0.08
Hen Dung (fresh)	1.63	1.54	0.85	0.24
Mixed Stable Manure	0.50	0.26	0.63	0.70

Table V. Average Composition of Fodders and Feeds.

Kind of Feeding Stuff.	Water.	Crude Fat.	Crude Fibre.	Crude Protein.	Crude Ash.	Carbohydrates.
GREEN FODDERS AND ENSILAGE.						
Pasture Grass	70.3	1.2	6.5	4.7	2.8	14.5
Orchard Grass (in bloom)	73.0	0.9	8.2	2.6	2.0	13.3
Timothy	61.6	1.2	11.8	3.1	2.1	20.2
Corn (Maize) Fodder—						
Flint varieties	79.8	0.7	4.3	2.0	1.1	12.1
Dent varieties	79.0	0.5	5.6	1.7	1.2	12.0
Sweet varieties	79.1	0.5	4.4	1.9	1.3	12.8
Red Clover	70.8	1.1	8.1	4.4	2.1	13.5
Alsike Clover (in bloom)	74.8	0.9	7.4	3.9	2.0	11.0
Alfalfa (Lucerne)	71.8	1.0	7.4	4.8	2.7	12.3
Crimson Clover (just heading)	89.2	0.4	1.8	2.5	1.2	4.9
Crimson Clover (full bloom)	81.5	0.6	5.1	3.2	1.5	8.1
Cow Pea	83.5	0.4	4.7	2.5	1.7	7.2
Sorghum (whole plant)	79.4	0.5	6.1	1.3	1.1	11.6
Rye Fodder	76.6	0.6	11.6	2.6	1.8	6.8
Oat Fodder	62.2	1.4	11.2	3.4	2.5	19.3
Corn (Maize) Ensilage	79.1	0.8	6.0	1.7	1.4	11.0
HAY AND DRY COARSE FODDERS.						
Corn (Maize) Fodder	42.2	1.6	14.3	4.5	2.7	34.7
Corn (Maize) Stalks	10.2	1.2	28.2	4.6	5.2	50.6
Hay, Mixed Meadow Grasses	16.0	2.1	29.9	6.4	4.6	41.0
Timothy Hay	13.6	2.5	28.9	5.9	4.4	44.7
Hay, Hungarian Grass	7.7	2.1	27.7	7.5	6.0	49.0
Red Clover Hay	15.3	3.3	24.8	12.3	6.2	38.1
Alsike Clover Hay	9.7	2.9	25.6	12.8	8.3	40.7
Alfalfa (Lucerne) Hay	8.4	2.7	25.0	14.3	7.4	42.7
Wheat Straw	9.6	1.3	38.1	3.4	4.2	43.4
Rye Straw	7.1	1.2	38.9	3.0	3.2	46.6
Oat Straw	9.2	2.3	37.0	4.0	5.1	42.4
ROOTS AND TUBERS.						
Mangels	90.9	0.2	0.9	1.4	1.1	5.5
Rutabagas	88.6	0.2	1.3	1.2	1.2	7.5
Turnips	90.5	0.2	1.2	1.1	0.8	6.2
Red Beets	88.5	0.1	0.9	1.5	1.0	8.0
Sugar Beets	86.5	0.1	0.9	1.8	0.9	9.8
Carrots	88.6	0.4	1.3	1.1	1.0	7.6
Potatoes	79.1	0.1	0.4	2.1	0.9	17.4
Sweet Potatoes	72.4	0.3	0.9	1.1	1.3	24.0

Table V. Average Composition of Fodders and Feeds.
(Concluded.)

KIND OF FEEDING STUFF.	POUNDS PER HUNDRED.					
	Water.	Crude Fat.	Crude Fibre.	Crude Protein.	Crude Ash.	Carbohydrates.
GRAINS AND OTHER SEEDS.						
Corn (Maize)—						
Flint	11.3	5.0	1.7	10.5	1.4	70.1
Dent	10.6	5.0	2.2	10.3	1.5	70.4
Sweet	8.8	8.1	2.8	11.6	1.9	66.8
Wheat (winter varieties)	10.5	2.1	1.8	11.8	1.8	72.0
Rye	11.6	1.7	1.7	10.6	1.9	72.5
Oats	11.0	5.0	9.5	11.8	3.0	59.7
Buckwheat	12.6	2.2	8.7	10.0	2.0	64.5
MILL PRODUCTS AND REFUSE FEEDS.						
Corn (Maize) Meal	14.4	3.8	1.9	9.3	1.4	69.2
Corn and Cob Meal	15.1	3.5	6.6	8.5	1.5	64.8
Corn Bran	8.5	8.1	11.5	11.4	0.8	59.7
Wheat Bran (all analyses)	11.7	4.1	8.9	15.4	5.9	54.0
Wheat Shorts	11.7	4.5	7.0	15.1	4.4	57.3
Wheat Middlings	11.8	4.0	4.4	15.7	3.2	60.9
Wheat Screenings	11.6	3.0	4.9	12.5	2.9	65.1
Rye Bran	11.6	2.8	3.4	14.4	3.5	64.3
Rye Shorts	9.3	2.8	5.1	18.0	4.9	59.9
Buckwheat Bran	12.9	5.9	13.4	22.1	4.3	41.4
Buckwheat Middlings	12.8	7.5	3.8	28.0	5.0	42.9
Rice Bran	9.7	8.8	9.5	12.1	10.0	49.9
Malt Sprouts	9.3	1.9	10.6	25.9	6.5	45.8
Brewers' Grains	75.7	1.7	3.7	5.9	0.9	12.1
Brewers' Grains, dried	8.7	6.6	13.1	22.7	3.8	45.1
Gluten Meal	8.0	14.6	1.6	33.0	1.3	41.5
Chicago Gluten Meal	9.1	5.5	1.3	33.7	0.9	49.5
Buffalo Gluten Feed	8.3	12.7	6.7	21.5	0.9	49.9
Grano Gluten Feed	6.0	14.2	11.4	31.0	2.7	34.7
Cerealine Feed	9.6	8.1	6.8	10.6	2.6	62.3
Hominy Chop	8.7	9.7	3.4	11.3	2.9	64.0
Corn Oil Meal	9.0	13.5	6.7	24.8	2.4	43.6
Cottonseed Meal	8.0	12.6	5.6	42.4	7.2	24.2
Linseed Meal (old process)	9.2	7.7	8.4	33.5	5.7	35.5
Linseed Meal (new process)	10.1	3.0	9.5	33.2	5.8	38.4

Table VI. Coefficients of Digestibility of American Feed Stuffs.

EXPERIMENTS WITH RUMINANTS.

KIND OF FODDER.	Crude Fibre or Cellulose.	Crude Fat.	Crude Protein.	Nitrogen-Free Extract. Carbohydrates.
HAY AND DRY COARSE FODDERS.				
Timothy Hay	53	61	48	63
Hay of Mixed Grasses (rich in protein)	60	49	59	59
Hay of Orchard Grass	61	55	60	55
Hay of Red Top	61	51	61	62
Dried Pasture Grass	77	60	72	73
Oat Straw	58	38	..	53
HAY OF LEGUMES.				
Cow-pea Vine Hay (fair quality)	43	50	65	71
Clover Hay (late bloom, fair quality)	46	53	55	64
Clover Hay (good quality)	48	43	49	58
Crimson Clover Hay	50	46	69	69
Alsike Clover	53	50	66	71
Alfalfa (Lucerne)	43	48	69	72
CORN FODDERS (PARTIALLY AIR DRY).				
Corn Stalks (whole plant)	67	52	52	64
Corn Stalks (leaves of)	61	63	56	50
Corn Fodder	65	74	59	74
Sweet Corn Fodder (mature)	74	74	64	68
GREEN FODDERS.				
Corn Fodder	52	76	53	74
Sweet Corn Fodder (milk)	75	74	77	81
Sorghum	59	74	46	74
Pasture Grass	76	63	70	73
Soiling Rye (formation of head)	80	74	79	71
Soiling Clover (late blossom)	53	65	67	78
Corn Ensilage	68	80	52	67
ROOTS, TUBERS, ETC.				
Potatoes	..	..	44	91
Sugar Beets	100	50	91	100
Mangels	43	..	75	91
GRAINS.				
Corn (Maize) Meal	..	92	60	93
Corn and Cob Meal	45	84	52	88
Pea Meal	26	55	83	94
Raw Cotton Seed	76	87	68	50
Soja-bean Meal	71	86	91	76

Table VI. Coefficients of Digestibility of American Feed Stuffs. — *Concluded.*

EXPERIMENTS WITH RUMINANTS.

KIND OF FODDER.	Crude Fibre or Cellulose.	Crude Fat.	Crude Protein.	Nitrogen-Free Extract, Carbohydrates.
BY-PRODUCTS.				
Cottonseed Meal	32	93	88	64
Chicago Gluten Meal	22	97	90	97
Gluten Meal	..	93	90	91
Buffalo Gluten Feed	100	94	89	89
Chicago Maize Feed	82	92	85	88
Winter Wheat Bran	28	65	78	71
Spring Wheat Bran	24	76	80	70
Wheat Middlings	36	85	85	88
New-process Linseed Meal	74	94	85	86
Old-process Linseed Meal	57	89	89	78
Malt Sprouts	34	100	80	69
Brewers' Grains Dried	53	91	79	59

Fuel Value of Food.

The different classes of food compounds or nutrients in a feed, in addition to their special functions of forming protein and fat, yield energy in the form of heat and muscular strength. The fuel or heat value of these nutrients has been measured and is expressed in calories.

A calorie is the amount of heat necessary to raise the temperature of a pound of water four degrees Fahrenheit. The calories in each of the three classes of nutrients are, on the average : —

	CALORIES.
In one pound of protein	1,860
In one pound of fats	4,220
In one pound of carbohydrates	1,860

The calories in a pound of fats are equal to those in about two and one-quarter pounds of protein or carbohydrates. In calculating the nutritive ratio of a ration, the fats are multiplied by two and one-quarter, in order to convert them into terms of carbohydrates. The digestible nutrients in the standard ration for milch cows contain 29,600 calories.

Table VII. Feeding Standards.

POUNDS PER DAY PER 1,000 POUNDS LIVE WEIGHT.

KIND OF ANIMAL.	Total dry Organic Matter.	Digestible Nutrients. Protein.	Carbohydrates, including Fibre.	Fat.	Total Nutritive Substances.	Nutritive Ratio.
Horse, at light work	21.0	1.5	9.5	0.40	11.40	1 : 7.0
" average work	22.5	1.8	11.2	0.60	13.60	1 : 7.0
" hard work	25.5	2.8	13.4	0.80	17.00	1 : 5.5
Oxen, at rest in stall	17.5	0.7	8.0	0.15	8.85	1 : 12.0
" ordinary work	24.0	1.6	11.3	0.30	13.20	1 : 7.5
" hard work	26.0	2.4	13.2	0.50	16.10	1 : 6.0
Oxen, fattening, first period	27.0	2.5	15.0	0.50	18.00	1 : 6.5
" " second period	26.0	3.0	14.8	0.70	18.50	1 : 5.5
" " third period	25.0	2.7	14.8	0.60	18.10	1 : 6.0
Milch Cows	24.0	2.5	12.5	0.40	15.40	1 : 5.4
Sheep, wool-producing (coarser breeds)	20.0	1.2	10.3	0.20	11.70	1 : 9.0
" " (finer breeds)	22.5	1.5	11.4	0.25	13.15	1 : 8.0
" fattening, first period	26.0	3.0	15.2	0.50	18.70	1 : 5.5
" " second period	25.0	3.5	14.4	0.60	18.50	1 : 4.5
Swine, fattening, first period	36.0	5.0	27.5		32.50	1 : 5.5
" " second period	31.0	4.0	24.0		28.00	1 : 6.0
" " third period	23.5	2.7	17.5		20.20	1 : 6.5
GROWING CATTLE. *Age, Months.* *Average live weight per head.*						
2–3 . . . . . 150 pounds	22.0	4.0	13.8	2.0	19.8	1 : 4.7
3–6 . . . . . 300 "	23.4	3.2	13.5	1.0	17.7	1 : 5.0
6–12 . . . . . 500 "	24.0	2.5	13.5	0.6	16.6	1 : 6.0
12–18 . . . . . 700 "	24.0	2.0	13.0	0.4	15.4	1 : 7.0
18–24 . . . . . 850 "	24.0	1.6	12.0	0.3	13.9	1 : 8.0

Table VIII. Fertilizer Constituents in Fodders and Feeds.

KIND OF FEEDING STUFF.	POUNDS PER TON.		
	Nitrogen.	Phosphoric Acid.	Potash.
Corn Fodder	14.4	6.8	11.2
Corn Stalks	14.6	4.6	19.8
Timothy Hay	18.4	6.6	28.4
Red Clover Hay	39.6	7.2	42.0
Scarlet Clover Hay	55.8	13.9	44.2
Alsike Clover Hay	41.0	13.4	44.6
Wheat Straw	10.0	1.8	14.4
Rye Straw	9.6	5.8	15.8
Oat Straw	12.8	4.4	24.2
Corn Kernels, Flint	33.6	14.0	8.0
Corn Kernels, Dent	33.0	14.0	8.0
Winter Wheat	37.8	18.6	12.8
Rye	34.0	17.0	11.2
Oats	37.8	17.8	13.4
Buckwheat	32.0	9.0	4.2
Wheat Bran	49.2	57.8	32.2
Wheat Middlings	50.2	28.2	14.0
Rye Bran	46.0	32.0	19.2
Corn Bran	36.4	3.8	1.4
Buckwheat Bran	70.8	34.0	22.8
Buckwheat Middlings	89.6	44.2	23.0
Malt Sprouts	82.8	32.6	37.0
Brewers' Grains	18.8	6.2	1.0
Brewers' Grains, Dried	72.0	21.8	1.6
Gluten Meal	105.0	11.2	1.4
Chicago Gluten Meal	110.4	5.8	1.0
Buffalo Gluten Feed	68.8	7.6	1.4
Cerealine Feed	33.8	25.0	13.4
Hominy Crop	36.2	29.8	14.6
Corn Oil Meal	79.2	29.0	3.4
Cottonseed Meal	135.4	61.6	38.0
Linseed Meal (old process)	107.2	38.6	28.2
Linseed Meal (new process)	106.2	35.6	27.2

INDEX.

Aberdeen Angus cattle, 179.
Air, food derived from, 10, 12.
Albuminoids, 139, 140.
Ammonia, definition of, 69; sulphate of, uses and composition, 72.
Analysis of soils, value of, 31.
Animal body, ash constituents of, 140; composition of, 137; functions of, 144; nitrogenous constituents of, 139.
Animal bone, 80.
Animal charcoal, 84.
Animal food, chemical analysis of, 143; classes of, 140.
Apatite, 87.
Ash, determination of, in feeds, 144; of animal body, 140.
Assimilation, 12.
Atavism in breeding, 171.
Atmosphere, action of, on soils, 20; as source of food, 12.
Ayrshire cattle, 178.
Azotine, 75.

Bacteria, effect of, on milk, 188.
Basic slag, 88.
Beet family, 128.
Blood, dried, 74.
Bone, composition of animal, 81; boiled, 82; fineness of, 82; raw, 82; steamed, 82.
Bone ash, 85.
Bone black, 84; superphosphate, 91.
Bran, as a feed, 150.
Breathing, functions of, 145.
Breed, influence of, on milk, 184; value of pure, 174.
Breeding, as a business, 174; atavism in, 171; cross, 173, 175; heredity in, 170; in-and-in, 173; lineage in, 172; prepotency in, 172; principles of, 170; variation in, 171.

Brewers' grains, as a feed, 150.
Buckwheat, as catch crop, 46.
Butter, color of, 193; composition of, 191; flavor of, 192; keeping quality of, 192; sweet-cream, 193; texture of, 192; yield of, 191.
Butter-fat, composition of, 182.
Buttermilk, composition of, 195.

Canadian apatite, 87.
Capillary attraction, 49.
Carbohydrates, composition of, 141; determination of, 144.
Carrot family, 128.
Castor pomace, 78.
Catch crops, 46.
Cattle, breeds of, beef, 179; butter, 177; milk, 177.
Cellulose, 144.
Cereals, description of, 129; as feeds, 149.
Charcoal, animal, 84.
Cheese, composition of, 193; as food, 194; manufacture of, 193.
Clay, composition and properties of, 24.
Clay soils, 26.
Claying of soils, 44.
Climate, effect of, 37.
Clovers, description of, 127, 132.
Colostrum, 186.
Commercial values, 106.
Composts, 59.
Cotswold sheep, 180.
Cottonseed meal, as feed, 152; as manure, 77.
Cow manure, 54.
Cream, composition of, 189.
Creaming, by centrifugal force, 190; by setting, 190.
Crops, demand for special, 113.
Crude fibre, determination of, 144.
Cultivating, 50.

Dairy cattle, breeds of, 177.
Dairying, elements of success in, 195, 196, 197.
Diffusion, 13.
Digestibility of feeds, 154.
Digestion coefficients, 155; function of, 145.
Drainage, function of, 42; methods of, 43.
Dried blood, 74.
Dried fish, 76.
Dried meat, 75.

Earth worms, effect on soils of, 22.
Egg-plants, 127.
Ensilage, 149.
Excretion, functions of, 146.

Fallow, bare, 116; cropping, 116.
Farming, extensive, 120; intensive, 120.
Farmyard manure, 53.
Fat, in animal food, 141; determination of, 143.
Feed, definition of, 146; digestibility of, 154; gluten, 151; manurial value of, 166, 168; mill, 150, 166.
Feeding, economy in, 164, 165; fertility in, 166; objects of, 157.
Feeding standards, 158; use of, 159.
Felt waste, 61.
Fermentation of manures, 57.
Ferments, effect of, on milk, 188.
Fertility of soils, 30; true measure of, 34.
Fertilizers, advantages of different, 103; analysis of, 108; complete, 102; formulas, 108; incomplete, 102; methods of buying, 102; special, 110; use of, 110.
Fertilizing elements, essential, 52.
Fertilizing materials, classes of, 68; standard, 100.
Fish, dried, 76.
Florida phosphate, 86.
Fodder, definition of, 146; green, 149.
Food, classes of animal, 140; definition of, 146.
Fruit crops, 135; manures for, 136; soils adapted to, 136.

Galloway cattle, 179.
Gas lime, 65.
Gelatinoids, 139.

Germination, 15; conditions necessary for, 16.
Gluten feeds, 151.
Grades, breeding of, 173.
Grasses, description of, 126, 131.
Green manuring. *See Manures.*
Guanos, phosphatic, 88.
Guaranteed composition, 104; interpretation of, 104.
Guernsey cattle, 177.
Gypsum, 66.

Hair waste, 61.
Harrowing, 50.
Hay, cutting of, best time for, 147; as a feed, 147.
Heredity in breeding, 170.
Hereford cattle, 179.
Holstein-Friesian cattle, 178.
Hominy meal, as a feed, 152.
Hoof meal, 77.
Horn meal, 77.
Horned Dorset sheep, 180.
Horny matter in animal body, 139.
Horse manure, 54.
Horses, breeds of; American trotting, 176; draft, 175; heavy carriage, 176; saddle, 176; thorough-bred, 176.
Humus, composition and effect of, 25.

Iron phosphate, 88.
Irrigation, 43.

Jersey cattle, 177.

Kainit, composition of, 97; use of, as manure, 97.

Lactation, period of, 186.
Land plaster, 66.
Leaf, structure of the, 12.
Leather meal, 61, 77.
Legumes, 127.
Leicester sheep, 180.
Lime, effect of, 24, 47, 51, 64, 67; gas, 65; shell, 65; slaked, 65.
Limestone, 65.
Limestone soils, 27.
Lincoln sheep, 180.
Lineage in breeding, 172.
Linseed meal, as feed, 152.
Loamy soils, 27.

INDEX.

Magnesia, sulphate of potash and, 99.

Manure, agricultural value of, 71; artificial, 68; care of, 56; commercial value of, 106; cow, 54; definition of, 52; farmyard, 53; application of, 58; for fruit crops, 136; green, 44, 47; crops useful as, 44; care in use of, 46; horse, 54; improvement of, 57; loss in, 55; natural, 53; nitrogenous, 69, 111; phosphatic, 80, 111; pig, 54; potassic, 95, 111; poultry, 58; preservers of, 56; sheep, 54; stable, composition of, 54; use of, 111, 112.

Market garden crops, 135.
Marl, use and composition of, 44, 63.
Meat, dried, 75.
Melon family, 128.
Merino sheep, 179.
Middlings, as feed, 150.
Milk, changes in, 188; composition of, 182, 183, 185; effect of bacteria on, 188; influence of food on, 187; properties of, 182; variations due to time of drawing, 186; yield of, from different breeds, 185.
Muck, use of, 59, 60.
Muriate of potash, 98.

Nitrate of potash, composition of, 72.
Nitrate of soda, composition of, 71.
Nitrates, application of, 79; use of, 69.
Nitrification in soils, 40.
Nitrogen, forms of, 69; as manure, 69, 78; organic, definition of, 69; uses of, 73, 79.
Nutritive ratio, 160.

Oyster shells, 65.

Palatability of rations, 162.
Peanut meal, as feed, 153.
Peat, use of, 59, 60.
Peaty soils, 27.
Pedigree, 173.
Phosphates, animal, 80; composition of, 90; definition of, 80; fixation of, 94; Florida, 86; insolubility of, 89; iron, 88; mineral, 85; odorless, 88; South Carolina, 86.
Phosphatic guanos, 88.

Phosphoric acid, insoluble, 89; in soils, 29; soluble, 91, 93.
Pig manure, 54.
Plant-food, sources of, 10.
Plant-food constituents, 9, 11; determination of, 11; functions of, 15; supply of, 14.
Plants, agricultural classification of, 128; air-dry, 8; annual, 16; biennial, 16; botanical classification of, 126; development of, 17; dry matter of, 8; life of, duration of, 16; parts of, 7; perennial, 17; water in, 7.
Plaster, land, 66; New York, 67; Nova Scotia, 66.
Plowing, fall, 48; methods of, 48; subsoil, 48.
Pony breeds, 176.
Potash, double sulphate of, and magnesia, 99; muriate of, 98; nitrate of, 72; in soils, 30; sulphate of, 98.
Potash manures, 95; forms of, 96.
Potash salts, appearance of, 99; uses of, 99.
Potatoes, sweet, 127; white, 127.
Poultry manure, 58.
Prepotency in breeding, 172.
Protein, determination of, 143.

Quick-lime, 65.

Rations, balanced, 164; examples of good, 163; preparation of, 161; wide *vs.* narrow, 161.
Respiration, functions of, 145.
Reversion in breeding, 171.
Rice bran, as feed, 153.
Rolling, 50.
Root crops, 133.
Roots, structure of, 13; functions of, 13.
Rotations, advantages of, 114; in dairy farms, 120; examples of good, 117, 118; for hay crops, 120; in market gardening, 120.
Rose family, 128.
Rye, as catch crop, 46.

Salt, use of, 67.
Saltpetre, 72.

Sand, composition and properties of, 23.
Sandy soils, 26.
Seed, adulteration of, 123; change of, 124; germinating power of, 125; what is good, 122; impurities in, 122; quality of, 123; selection of, 122; testing of, 125.
Sheep, breeds of, 179.
Sheep manure, 54.
Shorthorn cattle, 178.
Skim-milk, composition of, 194; as a food, 195.
Soils, absorptive properties of, 38; alluvial, 23; analysis of, 31; changes in, 37; classification of, 23; chemical composition of, 31; chemical improvement of, 50; clay, 26; claying of, 44; constituents of, 33; definition of, 18; drift, 23; effect of atmosphere on, 20; effect of lime on, 24, 47, 51, 64, 67; effect of growth of plants on, 21; effect of water on, 20; exhaustion of, 35; natural fertility of, 30; food obtained from, 13; formation of, 20; imperfections of, 41; improvement of, 41, 50; inorganic substances in, 29; limestone, 27; loamy, 27; marling of, 44; movement of, 22; nitrification in, 40; organic substances in, 29; origin of, 18; peaty, 27; perfect, 27; preparation of, 50; sandy, 26; sedentary, 22; natural strength of, 35; texture of, 36; transported, 23; vegetable, 27; weight of, 31.

South Carolina rock, 86; superphosphate, 91.
Southdown sheep, 180.
Stable manure, composition of, 54.
Stalks, as feed, 148.
Straw, as feed, 148.
Subsoil, definition of, 19; function of, 20.
Sulphate, of ammonia, 72; of potash, 98; of potash and magnesia, 99.
Superphosphates, composition of, 92; definition of, 90; use of, 95.
Swine, breeds of, 180.
Sylvinit, composition of, 97; use of, as manure, 97.

Tankage, 75.
Thomas phosphate meal, 88.
Tillage, 47.
Tomatoes, 127.
Tuber crops, 134; as food, 149.
Turnip family, 127.

Unit System, 106.

Values, commercial, 106.
Vegetable soils, 27.

Wastes, utilization of, 61.
Water, action of, on soils, 20.
Wheat plant, composition of, 32.
Wheat soil, composition of, 32.
Whey, 195.
Wolff's feeding standards, 158.
Wood ashes, use and composition of, 62.
Wool waste, 61.

www.ingramcontent.com/pod-product-compliance
Lightning Source LLC
Chambersburg PA
CBHW031816220426
43662CB00007B/677